P9-AQJ-475

DATE DUE

County Library System, FL - Blake Library

EMCO 38-296   MAR 1 7 2003

Newnes
Instrumentation and Measurement
Pocket Book
Third Edition

Newnes

Instrumentation and Measurement Pocket Book

Third edition

W. Bolton

 Newnes

OXFORD   AUCKLAND   BOSTON
JOHANNESBURG   MELBOURNE   NEW DELHI

Newnes
An imprint of Butterworth-Heinemann
Linacre House, Jordan Hill, Oxford OX2 8DP
225 Wildwood Avenue, Woburn, MA 01801-2041
A division of Reed Educational and Professional Publishing Ltd

A member of the Reed Elsevier plc group

First Published 1991
Second edition 1996
Reprinted 1998 (twice), 2000

© W. Bolton 1991, 1996

All rights reserved. No part of this publication may be
reproduced in any material form (including photocopying
or storing in any medium by electronic means and whether
or not transiently or incidentally to some other use of
this publication) without the written permission of the
copyright holder except in accordance with the provisions
of the Copyright, Designs and Patents Act 1988 or under
the terms of a licence issued by the Copyright Licensing
Agency Ltd, 90 Tottenham Court Road, London, England W1P
0LP. Application for the copyright holder's written permission
to reproduce any part of this publication should be
addressed to the publishers

British Library Cataloguing in Publication Data
A catalogue record for this book is available from the British
Library

Library of Congress Cataloguing in Publication Data
A catalogue record for this book is available from the Library of
Congress

ISBN 0 7506 2885 5

Typeset by Vision Typesetting, Manchester
Printed and bound by Antony Rowe Ltd., Chippenham and Reading

FOR EVERY TITLE THAT WE PUBLISH, BUTTERWORTH-HEINEMANN
WILL PAY FOR BTCV TO PLANT AND CARE FOR A TREE.

Contents

Part Four Microprocessor based systems

Preface to third edition

Aims: The main aims of this book are to:

- enable the reader to understand the 'jargon' of manufacturers' catalogues and data sheets for instruments and measurement systems;
- provide the reader with a basic understanding of the different types of instrumentation and measurement systems used for the measurement of quantities commonly encountered in engineering and process plant;
- provide a handy reference to a wide range of instrumentation and measurement systems;
- enable the reader to make an intelligent selection, and appreciate the limitations, of instrumentation and measurement systems.

Audience: The book is seen as being of relevance to:

- students in electrical, electronic, instrumentation, plant process and production engineering at craft, technician and undergraduate levels;
- technicians and engineers who use instrumentation in industry.

Structure of the book: The book is in four sections:

Part One: Systems
> This is concerned with the general form of measurement systems and their characteristics; this includes performance terminology, errors, dynamic characteristics, loading effects, noise and reliability.

Part Two: System components
> This deals with the components used to form measurement systems, namely transducers, signal converters and display systems.

Part Three: Measurements
> This part gives details of instruments and measurement systems for specific types of measurements; the measurements covered are those concerned with chemical composition, density, displacement, electrical quantities, flow, force, level, pressure, radiation, stress and strain, temperature and vacuum.

Part Four: Microprocessor based systems
> This is a look at microprocessor based instrumentation and deals with the basic elements of such systems and their interfacing with peripheral systems.

The emphasis throughout is on the characteristics of instruments and measurement systems. References are given for extension reading beyond the details given.

Changes for third edition: For the third edition, the section on microprocessor based instrumentation has been extended to give more details of microprocessor systems and interfacing.

W. Bolton

Part One
Systems

1 Measurement systems

The general measurement system

In general, measurement systems can be considered to have three basic constituent elements.

1 The *sensing element* or, as it is frequently called, the *transducer* is the element that produces a signal which is related to the quantity being measured. Such elements take information about the thing being measured and change it into some form which enables the rest of the measurement system to give a value to it.
2 The *signal converter* takes the signal from the sensing element and converts it into a condition which is suitable for the display part of a measurement system, or for use in a control system. The signal converter can be composed of three subelements, a *signal conditioner* which converts the signal from the sensing element into a physical form suitable for the display, a *signal processor* which improves the quality of the signal, e.g. amplifies it, and a *signal transmitter* to convey the signal some distance to the display.
3 The *display* element is where the output from the measuring system is displayed. The display element takes the information from the signal converter and presents it in a form which enable an observer to recognise it, e.g. a pointer moving across a scale.

The general form of a measurement system is thus a transducer connected to a signal converter which in turn is connected to a display element. It can be represented by a block diagram of the form shown in Figure 1.1.

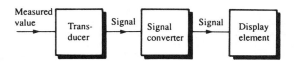

Figure 1.1 The general form of measurement system

System transfer function

For steady state conditions, the transfer function of a system is the ratio output θ_o to input θ_i (see page 24 for a general definition).

$$\text{Transfer function } G = \frac{\theta_o}{\theta_i}$$

A measurement system, however, can be made up of a transducer, signal conditioner and display (Figure 1.2). Each of these elements has its own transfer function. Thus for the transducer, transfer function G_1, with an input of θ_i and an output to the signal conditioner of θ_1:

$$G_1 = \frac{\theta_1}{\theta_i}$$

The signal conditioner, transfer function G_2, has an input of θ_1 and an output of θ_2. Thus

$$G_2 = \frac{\theta_2}{\theta_1}$$

Figure 1.2 Transfer function for a measurement system

The display, transfer function G_3, has an input of θ_2 and an output of θ_o. Thus

$$G_3 = \frac{\theta_o}{\theta_2}$$

The transfer function of the measurement system can be written as

$$G = \frac{\theta_o}{\theta_i} = \frac{\theta_1}{\theta_i} \times \frac{\theta_2}{\theta_1} \times \frac{\theta_o}{\theta_2}$$

$$G = G_1 \times G_2 \times G_3$$

The transfer function of the system is equal to the transfer function of the transducer multiplied by the transfer function of the signal conditioner multiplied by the transfer function of the display. If the system contained more elements, then provided the output signal from one element is the sole input to the next, the transfer function of the system is the product of the transfer functions of each of the elements.

Intelligent instruments

The term *intelligent* when applied to measurement systems means that a microprocessor or computer is included in the system. The term *dumb* is used when no such microprocessor is used. With a dumb instrument the system only gives a measure of a quantity, the observer has then to process and interpret the displayed data. With an intelligent instrument the measurement is made and then further processing occurs and the data is interpreted. Intelligent instruments can make decisions based on measurements made earlier, carry out calculations on data, manipulate information and initiate action based on the results obtained.

Further reading: Barney, G. C. (1988), *Intelligent Instrumentation*, Prentice Hall.

Calibration

Calibration is the process of putting marks on a display or checking a measuring system against a standard when the transducer is in a defined environment.

The basic standards from which all others derive are the *primary standards*. These are defined by international agreement and are maintained by national establishments, e.g. the National Physical Laboratory in Great Britain and the National Bureaux of Standards in the United States. There are seven such primary standards, and two supplementary ones. The seven are:

1 *Mass*. The kilogram is defined as being the mass of an alloy cylinder (90% platinum–10% iridium) of equal height and diameter, held at the International Bureau of Weights and Measures at Sèvres in France. Duplicates of this standard are held in other countries.

2 *Length*. The metre is defined as the length of path travelled by light in a vacuum during a time interval of $1/299\,792\,458$ of a second.

3 *Time*. The second is defined as a duration of 9 192 631 770 periods of oscillation of the radiation emitted by the caesium-133 atom under precisely defined conditions of resonance.

4 *Current*. The ampere is defined as that constant current which, if maintained in two straight parallel conductors of infinite length, of negligible circular cross-section, and placed one metre apart in a vacuum, would produce between these conductors a force equal to 2×10^{-7} N per metre of length.

5 *Temperature*. The kelvin (K) is defined so that the temperature at which liquid water, water vapour and ice are in equ'librium (known as the triple point) is 273.16 K.

6 *Luminous intensity*. The candela is defined as the luminous intensity, in a given direction, of a specified source that emits monochromatic radiation of frequency 540×10^{12} Hz and that has a radiant intensity of 1/683 watt per unit steradian (a unit solid angle).

7 *Amount of substance*. The mole is defined as the amount of a substance which contains as many elementary entities as there are atoms in 0.012 kg of the carbon-12 isotope.

The two supplementary standards are:

1 *Plane angle*. The radian is the plane angle between two radii of a circle which cuts off on the circumference an arc with a length equal to the radius.

2 *Solid angle*. The steradian is the solid angle of a cone which, having its vertex in the centre of the sphere, cuts off an area of the surface of the sphere equal to the square of the radius.

Primary standards are used to define national standards, not only in the primary quantities but also in other quantities which can be derived from them. For example, a resistance standard of a coil of manganin wire is defined in terms of the primary quantities of length, mass, time and current. Typically these national standards in turn are used to define reference standards which can be used by national bodies for the calibration of standards which are held in calibration centres. These centres then use their standards to carry out calibrations for industry. In a company there may well be such calibrated standards kept for checking the calibration of instrumentation in day-to-day use.

Table 1.1 lists some commonly used quantities and their relationship with the primary standards.

Table 1.1 Derived units

Quantity	Unit name	Unit in terms of primary units
Acceleration	metre per second squared	$m\,s^{-2}$
Angular acceleration	radian per second squared	$rad\,s^{-2}$
Angular velocity	radian per second	$rad\,s^{-1}$
Area	square metre	m^2
Capacitance	farad	$s^4A^2kg^{-1}m^{-2}$
Density	kilogram per cubic metre	$kg\,m^{-3}$
Electric charge	coulomb	$A\,s$

continued

Table 1.1 (*continued*)

Quantity	Unit name	Unit in terms of primary units
Electric field strength	volt per metre	$m\,kg\,A^{-1}\,s^{-3}$
Electric potential	volt	$m^2\,kg\,s^{-3}\,A^{-1}$
Energy	joule	$m^2\,kg\,s^{-2}$
Force	newton	$m\,kg\,s^{-2}$
Frequency	hertz	s^{-1}
Inductance	henry	$m^2\,kg\,s^{-2}\,A^{-2}$
Magnetic field strength	ampere per metre	$A\,m^{-1}$
Magnetic flux	weber	$m^2\,kg\,A^{-1}\,s^{-2}$
Magnetic flux density	tesla	$kg\,A^{-1}\,s^{-2}$
Power	watt	$m^2\,kg\,s^{-3}$
Pressure	pascal	$kg\,m^{-1}\,s^{-2}$
Resistance	ohm	$m^2\,kg\,A^{-2}\,s^{-3}$
Specific heat capacity	joule per kilogram kelvin	$m^2\,K^{-1}\,s^{-2}$
Speed	metre per second	$m\,s^{-1}$
Thermal conductivity	watt per metre kelvin	$m\,kg\,K^{-1}\,s^{-3}$
Volume	cubic metre	m^3

2 Performance terminology

The following are terms used generally in describing the performance of measurement systems or elements of such systems.

Accuracy. The accuracy of an instrument is the extent to which the reading it gives might be wrong. The term static accuracy is used when the quantity being measured is either not changing or changing very slowly, dynamic accuracy when it is changing quickly. Accuracy may be quoted as plus or minus some value of the variable, e.g. an ammeter might be quoted as having an accuracy of ± 0.1 A at some particular current value or for all its readings. An alternative is to quote accuracy as a *percentage of the full-scale deflection* (f.s.d.) of the instrument, e.g. an ammeter might have an accuracy quoted as $\pm 2\%$ f.s.d. This means that the accuracy of a reading of the ammeter when used for any reading within the range 0–10 A is plus or minus 2% of 10 A, i.e. plus or minus 0.2 A.

Bandwidth. The bandwidth can be defined as the range of frequencies for which the transfer function is no less than 70.7% of its peak value G (Figure 2.1). The 70.7% of G is $G/\sqrt{2}$. An alternative way of expressing this is that the bandwidth is the range of frequencies for which the transfer function is within 3 dB (decibels) of its peak value. A change of 3 dB means a transfer function which changes by $1/\sqrt{2}$.

$$\text{Change in dB} = 20 \log_{10}\left(\frac{\text{value}}{\text{max. value}}\right) = 20 \log_{10}\left(\frac{1}{\sqrt{2}}\right) = -3$$

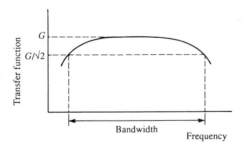

Figure 2.1 Bandwidth

Bel. See decibel.

Bias. The bias of an instrument is the constant error that exists for the full range of its measurements.

Binary word. A word is a grouping of a number of bits. With conventional or natural binary numbers then the position of the bits in a word has the significance that the least significant bit (LSB) is on the right end of the word and the most significant bit (MSB) on the left end. The denary value of the bits in a word is

$$2^{n-1} \ldots 2^4\ 2^3\ 2^2\ 2^1\ 2^0$$
MSB   LSB

Bit. This is the abbreviation for a binary digit, 0 or 1.

Cross-talk. This is the interference that occurs between neighbouring channels with a multiplexer or other element which has parallel inputs.

Dead space. The dead space of an instrument is the range of values of the quantity being measured for which it gives no reading.

Decibels. The ratio between two values of electric or acoustic power is usually expressed on a logarithmic scale. With base-10 logarithms the ratio is given the unit the bel. The decibel is one-tenth of a bel.

$$N_{bel} = \log_{10}(P_1/P_2)$$
$$N_{dB} = 10 \log_{10}(P_1/P_2)$$

When two voltages V_1 and V_2 operate into identical impedances, i.e. $P_1 = V_1^2/Z$ and $P_2 = V_2^2/Z$, then $P_1/P_2 = V_1^2/V_2^2$ and so

$$N_{dB} = 20 \log_{10}(V_1/V_2)$$

Similarly for the currents

$$N_{dB} = 20 \log_{10}(I_1/I_2)$$

Discrimination. The discrimination of an instrument is the smallest change in the quantity being measured that will produce an observable change in the reading of the instrument.

Drift. An instrument is said to show drift if there is a gradual change in output over a period of time which is unrelated to any change in input. See Zero drift.

Error. The error of a measurement is the difference between the result of the measurement and the true value of the quantity being measured.

    Error = measured value − true value

Gain. The gain of a system or element is the output divided by the input.

$$\text{Gain} = \frac{\text{Output}}{\text{Input}}$$

Hysteresis. Instruments can give different readings, and hence an error, for the same value of measured quantity according to whether that value has been reached by a continuously increasing change or a continuously decreasing change (Figure 2.2). This effect, called hysteresis, occurs as a result of such things as bearing friction and slack motion in gears in instruments. The hysteresis error is the difference

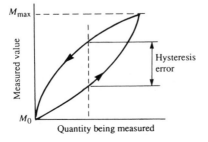

Figure 2.2 Hysteresis

between the measured values obtained when the measured quantity is increasing and when decreasing to that value. Hysteresis is often expressed in terms of the maximum hysteresis as a percentage of the full-scale-deflection.

$$\text{Hysteresis} = \frac{\text{Maximum hysteresis error}}{M_{max} - M_0} \times 100\%$$

Lag. A system is said to show lag if when the quantity being measured changes the measurement system does not respond instantaneously but some time later.

Non-linearity error. A linear relationship for an element or a system means the output is directly proportional to the input. In many instances however, though a linear relationship is used it is not perfectly linear and so errors occur. The non-linearity error is the difference between the true value and what is indicated when a linear relationship is assumed. Thus for Figure 2.3, the non-linearity error when the instrument indicates M is N. Non-linearity is often expressed in terms of the maximum non-linearity error as a percentage of the full-scale-deflection. Thus for Figure 2.3

$$\text{non-linearity} = \frac{N_{max}}{M_{max} - M_0} \times 100\%$$

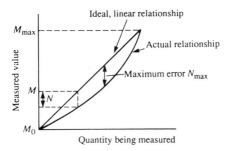

Figure 2.3 Non-linearity

Precision. Precision is a measure of the scatter of results obtained from measurements as a result of random errors. It describes the closeness of the agreement occurring between the results obtained for a quantity when it is measured several times under the same conditions.

Quantization error. This is the error obtained with a digital-to-analogue or analogue-to-digital converter due to the quantization interval, this being the contribution of the least significant bit.

Quantization noise. This is the noise that can be considered to be added to an analogue signal as a consequence of the quantization error.

Range. The range of an instrument is the limits between which readings can be made.

Reliability. The reliability of an instrument is the probability that it will operate to an agreed level of performance under the conditions specified for its use.

Repeatability. The repeatability of an instrument is its ability to display the same reading for repeated applications of the same value of the quantity being measured.

Reproducibility. The reproducibility of an instrument is its ability to display the same reading when it is used to measure a constant quantity over a period of time or when that quantity is measured on a number of occasions.

Resolution. The resolution of an instrument is the smallest change in the quantity being measured that will produce an observable change in the reading of the instrument.

Response time. When the quantity being measured changes, a certain time, called the response time, has to elapse before the measuring instrument responds fully to the change.

Sample rate. Some instruments, e.g. digital voltmeters, take samples of the variable at regular intervals. The greater the sample rate, i.e. the greater the number of samples taken per second, the more readily the instrument readings mirror a rapidly changing input.

Sensitivity. The sensitivity of an instrument is:

$$\text{sensitivity} = \frac{\text{change in instrument scale reading}}{\text{change in the quantity being measured}}$$

i.e. the rate of change of output of the system with respect to the input.

Sensitivity drift. The sensitivity drift is the amount by which the sensitivity changes as a result of changes in environmental conditions.

Signal to noise ratio. The signal to noise ratio is the ratio of the signal level V_s to the internally generated noise level V_n. It is usually expressed in decibels, i.e.

$$\text{signal to noise ratio in dB} = 20 \log_{10}(V_s/V_n)$$

Slew rate. The slew rate is the maximum rate of change with time that the output can have.

Span. The span of an instrument is the limits between which readings can be made.

Stability. The stability of an instrument is its ability to display the same reading when it is used to measure a constant quantity over a period of time or when that quantity is measured on a number of occasions.

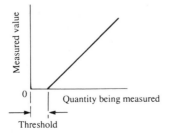

Figure 2.4 Threshold

Threshold. The threshold is the minimum value a signal must have reached before the instrument responds and gives a detectable reading (Figure 2.4). It is the dead space when the input starts from a zero value.

Time constant. A system when subject to an abrupt change in input, i.e. a step input, takes time to attain its final output (Figure 2.5). The time constant is the time taken for it to reach 63.2% of this final output. See Chapter 4.

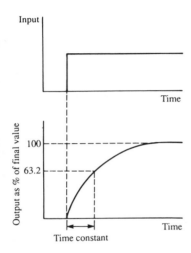

Figure 2.5 Time constant

Transfer function. The transfer function is the ratio of the output of a system, or an element of that system, to its input.

$$\text{Transfer function} = \frac{\text{output}}{\text{input}}$$

True value. This is the value with zero error.

Word. See binary word.

Zero drift. This term is used to describe the change in the zero reading of an instrument that can occur with time.

3 Errors

Sources of error

Errors can be classified as being either random or systematic errors. *Random errors* are ones which can vary in a random manner between successive readings of the same quantity. *Systematic errors* are errors which do not vary from one reading to another. The following are common sources of such errors with measurement systems.

1 Random errors

(a) *Operating errors.* These can result from a variety of causes, e.g. errors in reading the position of a pointer on a scale due to the scale and pointer not being in the same plane, the reading obtained then depending on the angle at which the pointer is viewed against the scale (the so-called parallax error). Also there are errors due to the uncertainty that exists in estimating readings between scale markings on an instrument's display.

(b) *Environmental errors.* These are errors which can arise as a result of environmental effects, such as a change in temperature or electromagnetic interference.

(c) *Stochastic errors.* These result from stochastic processes such as noise (see Chapter 6). A stochastic process is one which results in random signals.

2 Systematic errors

(a) *Construction errors.* These occur in the manufacture of an instrument and arise from such causes as tolerances on the dimensions of components and on the values of electrical components used.

(b) *Approximation errors.* These arise from assumptions made regarding relationships between quantities, e.g. a linear relationship between two quantities is often assumed and may in practice only be an approximation to the true relationship.

(c) *Ageing errors.* These are errors resulting from the instruments getting older, e.g. components deteriorating and their values changing, a build-up of deposits on surfaces affecting contact resistances and insulation.

(d) *Insertion errors.* These are errors which result from the insertion of the instrument into the position to measure a quantity affecting its value, e.g. inserting an ammeter into a circuit to measure the current thus changing the value of the current due to the ammeter's own resistance.

Spread of results

The results of a number of measurements of the same quantity can be plotted as frequency distribution. The frequency is the number of times a particular value, or values within a range of values, occurs. This is plotted against the values to give the frequency distribution (Figure 3.1). The distribution shows how the values obtained from the measurements vary. The greater the spread of the distribution the less the precision of the measurement.

In presenting a value which is representative of a number of measurements of a quantity the following are often used:

1   *Arithmetic mean* $(\bar{x})$.   This is the sum of all the results divided by the number n of results considered.

$$\bar{x} = \frac{\Sigma x}{n}$$

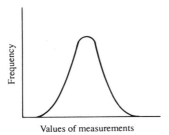

Figure 3.1 A frequency distribution

2 *Mode*. This is the value with the greatest frequency. If a distribution is symmetrical then the mean and the mode will have the same value. They will however differ if this is not the case (Figure 3.2).

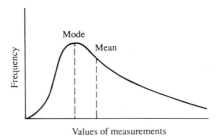

Figure 3.2 Mean and mode

3 *Median*. This is the value which divides the frequency distribution into two equal areas. In the case of a symmetrical distribution it will have the same value as the mean.

A measure of the precision, i.e. spread of a frequency distribution, is given by the *root mean square deviation* or *standard deviation*. The *deviation d* for a measurement is the difference between the mean and the value of that measurement. The sum of the squared deviations for all the measurements obtained (Σd^2) divided by the number of measurements n gives the mean of the squares of the deviations. The square root of this quantity is the root mean square deviation or standard deviation σ.

$$\sigma = \sqrt{\frac{\Sigma d^2}{n}}$$

Probable error

The frequency distribution of a set of measurements can be considered to show the deviations, i.e. errors, of the various measurements from the mean. The frequency distribution frequently takes the form shown in Figure 3.3, the form being referred to as a *Gaussian distribution*. Such a distribution shows that the most frequent measurement is at

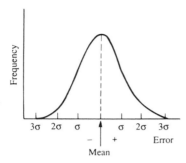

Figure 3.3 Gaussian distribution

the mean and has no error; small errors are more likely than large errors; and there is an equal chance of measurements having errors greater than the mean, plus errors, and error less than the mean, minus errors.

With the Gaussian distribution, the chance of a measurement occurring within one standard deviation of the mean is 68.3%, within two standard deviations 95.5%, three standard deviations 99.7%, four standard deviations 99.99%. The chance of a measurement falling within 0.6745σ qf the mean is 50%. The 0.6745σ is called the *probable error*.

Thus a statement of the probable error for a batch of components means that there is a 50% chance that if we take, at random, one of the components it will have a random error no greater than $\pm 0.6745\sigma$ from the mean value.

Limiting error

In the case of some components and instruments their deviations from the specified values is guaranteed to be within a certain percentage of that value. The deviations in this case are then referred to as *limiting errors*.

Summation of errors

A quantity may be determined as a result of calculations carried out on the results of a number of measurements, each of which has some error associated with it. When the result is obtained by:

1 Adding measurements: add the errors to obtain the overall error.
2 Subtracting measurements: add the errors to obtain the overall error.
3 Multiplying measurements: add the percentage errors to obtain the overall percentage error.
4 Dividing measurements: add the percentage errors to obtain the overall percentage error.
5 The measurement as a power: multiply the percentage error of the measurement by the power to obtain the overall percentage error.

The derivation of the above relationships can be illustrated by considering the addition of measurements. Suppose the quantity X is obtained by adding together the results of the measurements of two quantities A and B. Then in the absence of any errors in the measurements

$$X = A + B$$

However, taking into account errors,

$$X \pm \delta X = A \pm \delta A + B \pm \delta B$$

Thus

$$\delta X = \delta A + \delta B$$

The result of adding the two measurements is to add the errors.

When the quantity is obtained as a result of multiplying two measurements, then in the absence of errors

$$X = AB$$

Taking into account errors

$$X \pm \delta X = (A \pm \delta A)(B \pm \delta B)$$

Neglecting small quantities,

$$X \pm \delta X = AB \pm A\delta B \pm B\delta A$$

$$\delta X = A\delta B + B\delta A$$

Hence

$$\frac{\delta X}{X} = \frac{A\delta B + B\delta A}{AB}$$

$$\frac{\delta X}{X} = \frac{\delta B}{B} + \frac{\delta A}{A}$$

$$\frac{\delta X}{X} \times 100 = \frac{\delta B}{B} \times 100 + \frac{\delta A}{A} \times 100$$

The percentage error in X is the sum of the percentage errors in the measurements.

System accuracy

If the transfer function of the transducer is G_1, the input θ_i and the output signal θ_1, then in the absence of any error

$$\theta_1 = G_1\theta_i$$

Because of errors the output will fall within a range of values $(\theta_1 \pm \delta\theta_1)$, hence the transfer function G_1 must assume a spread of values and should thus be written as $(G_1 \pm \delta G_1)$. Hence the relationship between the input and output should be

$$\theta_1 \pm \delta\theta_1 = (G_1 \pm \delta G_1)\theta_i$$

The output signal from the transducer becomes the input signal to the signal conditioner. Because of errors this has a transfer function $(G_2 \pm \delta G_2)$ and gives an output signal $(\theta_2 \pm \delta\theta_2)$.

$$\theta_2 \pm \delta\theta_2 = (G_2 \pm \delta G_2)(\theta_1 \pm \delta\theta_1)$$
$$= (G_2 \pm \delta G_2)(G_1 \pm \delta G_1)\theta_i$$

The output signal from the signal conditioner becomes the input to the display. Because of errors this has a transfer function $(G_3 \pm \delta G_3)$ and gives an output signal $(\theta_o \pm \delta\theta_o)$.

$$\theta_o \pm \delta\theta_o = (G_3 \pm \delta G_3)(\theta_2 \pm \delta\theta_2)$$
$$= (G_3 \pm \delta G_3)(G_2 \pm \delta G_2)(G_1 \pm \delta G_1)\theta_i$$

θ_o is the output and $\delta\theta_o$ the error for the entire system for the input θ_i.

If small terms are ignored, then

$$\theta_o \pm \delta\theta_o = (G_3 G_2 G_1 \pm G_2 G_1 \delta G_3 \pm G_3 G_1 \delta G_2 \pm G_3 G_2 \delta G_1)\theta_i$$

$$= G_3 G_2 G_1 \left(1 \pm \frac{\delta G_3}{G_3} \pm \frac{\delta G_2}{G_2} \pm \frac{\delta G_1}{G_1}\right)\theta_i$$

In the absence of any errors the equation would be

$$\theta_o = G_3 G_2 G_1 \theta_i$$

Thus $G_1 G_2 G_3$ is the overall nominal gain of the system. Hence, dividing both sides of the equation by θ_o gives

$$1 \pm \frac{\delta\theta_o}{\theta_o} = 1 \pm \frac{\delta G_3}{G_3} \pm \frac{\delta G_2}{G_2} \pm \frac{\delta G_1}{G_1}$$

$$\frac{\delta\theta_o}{\theta_o} = \frac{\delta G_3}{G_3} + \frac{\delta G_2}{G_2} + \frac{\delta G_1}{G_1}$$

$\delta\theta_o/\theta_o$ is the error in the output as a fraction of the output. $\delta G/G$ is the error in a transfer function as a fraction of the transfer function. Hence the equation is just stating the fractional error in the output is the sum of the fractional errors in each element in the measuring system or, alternatively, the percentage error in the output is the sum of the percentage errors in each element.

4 Dynamic characteristics

Static and dynamic characteristics

The *static characteristics* of an instrument refer to the steady state reading that it gives when it has settled down. The *dynamic characteristics* describe the behaviour of an instrument in the time between when the measured quantity changes and a steady reading is given.

Zero order instruments

Instruments are said to be *zero order* when the output or reading is instantaneously reached following a change in the measured quantity (Figure 4.1). The relationship between the output θ_o and the input θ_i for such an instrument does not include any term involving time and is of the form:

$$\theta_o = k\theta_i$$

where k is a constant. An example of such an instrument is a potentiometer, the output voltage changing immediately the slider is moved along the potentiometer track.

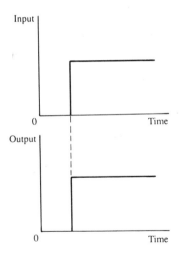

Figure 4.1 Response of a zero order system to a step input

First order instruments

Instruments are said to be *first order* when the relation between the input and output depends on the rate at which the output changes. For such a system the relationship between the input θ_i and the output θ_o is of the form

$$a_1 \frac{d\theta_o}{dt} + a_0 \theta_o = b_0 \theta_i$$

where $d\theta_o/dt$ is the rate at which the output changes and a_1, a_0 and b_0

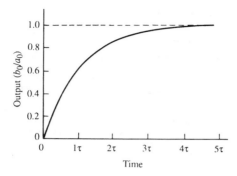

Figure 4.2 Response of a first order system to a step input

are constants. For an abrupt change in input, i.e. a step input, the output varies with time according to the relationship

$$\frac{\theta_o}{\theta_i} = \left(\frac{b_0}{a_0}\right)(1 - e^{-t/\tau})$$

where τ is a_1/a_0 and is known as the *time constant*. After 1τ the value of θ_o/θ_i is $0.63(b_0/a_0)$. After 2τ it is $0.87(b_0/a_0)$, after 3τ $0.95(b_0/a_0)$, after 4τ $0.98(b_0/a_0)$, after 5τ $0.99(b_0/a_0)$. The steady state value is thus (b_0/a_0). Figure 4.2 shows a graph of such an output change. A thermometer is an example of a first order instrument (see heat transfer effects later in this chapter).

Heat transfer effects

Consider a thermometer at temperature T in a liquid at temperature T_l.

Rate at which heat enters thermometer $\dfrac{dQ}{dt} = k(T_l - T)$

where k is a constant. If the thermometer has a specific heat capacity c and a mass m then a heat input of δQ in a time δt gives a temperature change δT where

$$\delta Q = mc\delta T$$

Hence

$$\frac{dQ}{dt} = mc\frac{dT}{dt}$$

$$k(T_l - T) = mc\frac{dT}{dt}$$

$$mc\frac{dT}{dt} + kT = kT_l$$

Thus for a step input, e.g. a thermometer at one temperature abruptly put into a liquid at another temperature,

$$\frac{T}{T_l} = (1 - e^{-t/\tau})$$

where the time constant $\tau = mc/k$.

Second order instruments

Instruments are said to be *second order* when the relation between the input θ_i and output θ_o is of the form

$$a_2 \frac{d^2\theta_o}{dt^2} + a_1 \frac{d\theta_o}{dt} + a_0\theta_o = b_0\theta_i$$

where b_0, a_0, a_1 and a_2 are constants. When the output has stopped changing the equation becomes $a_0\theta_o = b_0\theta_i$.

Such a system when subject to a step input can give an oscillating output. The *natural frequency* ω_n of the oscillation is given by

$$\omega_n = \sqrt{(a_0/a_2)}$$

The oscillation is subject to damping, the *damping ratio* ζ being given by

$$\zeta = \frac{a_1}{2\sqrt{(a_0 a_2)}}$$

The differential equation is thus usually written as

$$\frac{1}{\omega_n^2} \frac{d^2\theta_0}{dt^2} + \frac{2\zeta}{\omega_n} \frac{d\theta_o}{dt} + \theta_o = \frac{b_0}{a_0}\theta_i$$

When $\zeta = 1$, a condition known as *critical damping*, then

$$\theta_o/\theta_i = (b_0/a_0)[1 - (\exp - \omega_n t)(1 + \omega_n t)]$$

When ζ is less than 1 the system is said to be underdamped, when greater than 1, overdamped. Figure 4.3 shows graphs of how the output varies with time for a range of values of ζ.

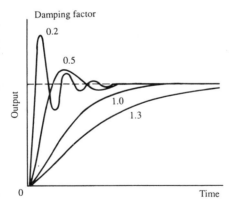

Figure 4.3 Responses of a second order system to a step input

Damped spring system

Many second order elements can be considered to be essentially a mass m, a spring and damping (Figure 4.4). The net force applied to the mass is the applied force F minus the force resulting from the stretching or compressing of the spring and minus the force from the damper. The force resulting from stretching or compressing the spring is propor-

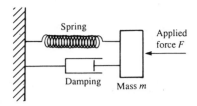

Figure 4.4 Mass, spring and damper system

tional to the change in length x of the spring, i.e. kx with k being the spring stiffness. The damper can be thought of as a piston moving in a container filled with oil. The force resulting from the damper displacement is proportional to the rate at which the displacement of the piston is changing, i.e. $c\,dx/dt$ where c is a constant. Thus

$$\text{net force applied to mass } m = F - kx - c\,\frac{dx}{dt}$$

This net force will cause an acceleration (Newton's second law). Acceleration is rate of change of velocity (dv/dt) and velocity is rate of change of displacement (dx/dt). Hence acceleration is (d^2x/dt^2). Hence

$$F - kx - c\,\frac{dx}{dt} = m\,\frac{d^2x}{dt^2}$$

$$m\,\frac{d^2x}{dt^2} + c\,\frac{dx}{dt} + kx = F$$

In the absence of the damping a mass m on the end of a spring would freely oscillate with a natural angular frequency ω_n given by

$$\omega_n = \sqrt{(k/m)}$$

The motion is damped with a damping ratio ζ defined as

$$\zeta = \frac{c}{2\sqrt{(mk)}}$$

Thus the equation becomes

$$\frac{1}{\omega_n^2}\,\frac{d^2x}{dt^2} + \frac{2\zeta}{\omega_n}\,\frac{dx}{dt} + x = \frac{F}{k}$$

D operator

With a differential equation $d\theta/dt$ can be replaced by $D\theta$, with D being called the operator. Similarly $d^2\theta/dt^2$ can be replaced by $D^2\theta$. In general

$$D^n\theta = \frac{d^n\theta}{dt^n}$$

The D, in combinations with constants and positive integral powers of itself, can be manipulated by the ordinary rules of algebra. Thus a first order differential equation

$$a_1\,\frac{d\theta_o}{dt} + a_0\theta_o = b_0\theta_i$$

becomes

$$a_1 D\theta_o + a_0\theta_o = b_0\theta_i$$

$$\frac{\theta_o}{\theta_i} = \frac{b_0}{a_1 D + a_0} = \frac{G}{\tau D + 1}$$

where $G = (b_0/a_0)$ and $\tau = a_1/a_0$.

There are a number of procedures that can be used to solve differential equations in this form. The following is one method. Consider the equation of the form

$$(a D^2 + b D + c)y = 0$$

If $y = Ae^{mx}$ then $Dy = Ame^{mx}$ and $D^2y = Am^2e^{mx}$ and this would mean that

$$a Am^2 e^{mx} + b Ame^{mx} + cy = 0$$
$$Ae^{mx}(am^2 + bm + c) = 0$$

Thus $y = Ae^{mx}$ can only be a solution provided $(am^2 + bm + c) = 0$. This equation is called the auxiliary equation. The roots of the equation can be obtained by factorising or using the formula

$$m = \frac{-b \pm \sqrt{(b^2 - 4ac)}}{2a}$$

If $b^2 > 4ac$ then there are two different real roots α and β and the general solution is

$$y = Ae^{\alpha x} + Be^{\beta x}$$

If $b^2 = 4ac$ there are two equal roots α and

$$y = (Ax + B)e^{\alpha x}$$

If $b^2 < 4ac$ there are two complex roots, $\alpha \pm j\beta$, and

$$y = e^{\alpha x}(A \cos \beta x + B \sin \beta x)$$

The constants A and B are determined using given boundary conditions.

Further reading: Distefano, J. D., Stubberud, A. R. and Williams, I. J. (1987), *Theory and Problems of Feedback and Control*, McGraw-Hill.

D operator and first order system

Consider the equation

$$\frac{\theta_o}{\theta_i} = \frac{1}{\tau D + 1}$$

$$(\tau D + 1)\theta_o = \theta_i$$

The auxiliary equation is formed by replacing the differential operator by an algebraic variable m and equating θ_i to zero. Hence

$$\tau m + 1 = 0$$

This means that $m = -1/\tau$. Hence the solution is of the form

$$\theta_o = Ae^{mt} = Ae^{-t/\tau}$$

This gives the transient solution.

Consider now a step input. When steady state conditions occur then the value of $D\theta_o$ is zero. This would mean that $\theta_o/\theta_i = 1/(0+1) = 1$. Hence

$$\theta_o = Ae^{-t/\tau} + \theta_i$$

For a step input we might have $\theta_o = 0$ when $t = 0$, then

$$0 = Ae^0 + \theta_i$$

and so $A = -\theta_i$. Hence the complete solution for a step input is

$$\theta_o = \theta_i(1 - e^{-t/\tau})$$

D operator and second order system

Consider the above procedure applied to a second order system of the form

$$(D^2 + 2\zeta\omega_n D + \omega_n^2)\theta_o = \omega_n^2\theta_i$$

The auxiliary equation is

$$m^2 + 2\zeta\omega_n m + \omega_n^2 = 0$$

This will have roots given by

$$m = \frac{-2\zeta\omega_n \pm \sqrt{(4\zeta^2\omega_n^2 - 4\omega_n^2)}}{2}$$

$$= -\zeta\omega_n \pm \omega_n\sqrt{(\zeta^2 - 1)}$$

When $\zeta > 1$ there are two real and unequal roots and so the transient change with time is given by

$$\theta_o = A\exp(m_1 t) + B\exp(m_2 t)$$

where $m_1 = -\zeta\omega_n + \omega_n\sqrt{(\zeta^2 - 1)}$ and $m_2 = -\zeta\omega_n - \omega_n\sqrt{(\zeta^2 - 1)}$. When $\zeta = 1$ there are two equal real roots and

$$\theta_0 = (At + B)\exp(-\zeta\omega_n t)$$

When $\zeta < 1$ there are two complex roots and

$$\theta_0 = \exp(-\zeta\omega_n t)(A\sin\omega t + B\cos\omega t)$$

where $\omega = \omega_n\sqrt{(1 - \zeta^2)}$.

With a step input of $\theta_i = 0$ at $t = 0$ and a steady state value of given by $D^2\theta_0 = 0$ and $D\theta_0 = 0$ then the constants can be evaluated and the complete solution for both the transient and steady state obtained.

Laplace transform

Laplace transforms enable differential equations to be transformed into equations which can then be handled as simple algebraic equations. The Laplace transform of some function of time $f(t)$ is defined by

$$\text{Laplace transform } F(s) = \int_0^\infty f(t)e^{-st}\,dt$$

where s is a complex quantity.

In using the Laplace transform to obtain the solution to a differential equation, the procedure to be adopted is:

1. Transform the differential equation into its Laplace equivalent.
2. Carry out all algebraic manipulations, e.g. consider what happens when a step input is applied to the system, and obtain a Laplace solution.
3. Convert the Laplace solution back into an equation giving a function of time, i.e. invert the Laplace transformation operation. In order to use tables of Laplace transforms to carry out the conversion, it is often necessary to first use partial fractions to put the Laplace solution into a suitable form.

The following are basic operations involved in transforming a differential equation into a Laplace form. It has been assumed that $f(t)$ has the value zero for all times before $t = 0$.

1   The addition of two functions becomes the addition of their two Laplace transforms.

$f_1(t) + f_2(t)$ becomes $F_1(s) + F_2(s)$

2   The subtraction of two functions becomes the subtraction of their two Laplace transforms.

$f_1(t) - f_2(t)$ becomes $F_1(s) - F_2(s)$

3   The multiplication of some function by a constant becomes the multiplication of the Laplace transform of the function by the same constant.

$af(t)$ becomes $aF(s)$

4   A function which is delayed by a time T, i.e. $f(t - T)$, becomes $e^{-Ts}F(s)$ for values of T greater than or equal to zero.

5   The first derivative of some function becomes s times the Laplace transform of the function minus the value of the function at $t = 0$

$$\frac{d}{dt}f(t) \text{ becomes } sF(s) - f(0)$$

6   The second derivative of some function becomes s^2 times the Laplace transform of the function minus the value of the function and s times its rate of change at $t = 0$

$$\frac{d^2}{dt^2}f(t) \text{ becomes } s^2F(s) - f(0) - s\frac{d}{dt}f(0)$$

7   The first integral of some function, between zero time and time t, becomes $(1/s)$ times the Laplace transform of the function.

$$\int_0^t f(t) \text{ becomes } \frac{1}{s}F(s)$$

Table 4.1 gives some of the more common Laplace transforms and their corresponding time functions.

Table 4.1   Laplace transforms

Laplace transform	Time function	
1		A unit impulse
$\dfrac{1}{s}$		A unit step function
$\dfrac{1}{s^2}$	t	A unit slope ramp function
$\dfrac{1}{s^3}$	$\dfrac{t^2}{2}$	

continued

Table 4.1 (*continued*)

Laplace transform	Time function	
$\dfrac{1}{s+a}$	e^{-at}	Exponential decay
$\dfrac{1}{(s+a)^2}$	te^{-at}	
$\dfrac{a}{s(s+a)}$	$1-e^{-at}$	Exponential growth
$\dfrac{a}{s^2(s+a)}$	$t-\dfrac{(1-e^{-at})}{a}$	
$\dfrac{s}{(s+a)^2}$	$(1-at)e^{-at}$	
$\dfrac{\omega}{s^2+\omega^2}$	$\sin \omega t$	Sine wave
$\dfrac{s}{s^2+\omega^2}$	$\cos \omega t$	Cosine wave
$\dfrac{\omega}{(s+a)^2+\omega^2}$	$e^{-at}\sin \omega t$	Damped sine wave
$\dfrac{s+a}{(s+a)^2+\omega^2}$	$e^{-at}\cos \omega t$	Damped cosine wave
$\dfrac{\omega^2}{s(s^2+\omega^2)}$	$1-\cos \omega t$	
$\dfrac{\omega^2}{s^2+2\zeta\omega s+\omega^2}$	$\dfrac{\omega}{\sqrt{(1-\zeta^2)}}e^{-\zeta\omega t}\sin[\omega\sqrt{(1-\zeta^2)}t]$	

Further reading: Distefano, J. D., Stubberud, A. R. and Williams, I. J. (1987), *Theory and Problems of Feedback and Control*, McGraw-Hill.

Laplace transform for a first order system

For a first order element the differential equation is of the form

$$a_1 \frac{d\theta_o}{dt} + a_0\theta_o = b_0\theta_i$$

The corresponding Laplace transform if $\theta_o = 0$ at $t-0$

$$a_1 s \times \theta_o(s) + a_0 \times \theta_o(s) = b_0 \times \theta_i(s)$$

The transfer function $G(s)$ is defined as the ratio of the Laplace transform of the output $\theta_o(s)$ to the Laplace transform of the input $\theta_i(s)$. Thus

$$G(s) = \frac{b_0}{a_1 s + a_0}$$

This can be rearranged to give

$$G(s) = \frac{b_0/a_0}{(a_1/a_0)s + 1}$$

b_0/a_0 is the steady state transfer function G of the system. a_1/a_0 is the time constant τ of the system. Hence

$$G(s) = \frac{G}{\tau s + 1}$$

Consider the behaviour of a first order system when subject to a step input. The Laplace transform of the output is thus

Laplace transform of output $= G(s) \times$ Laplace transform of input

The Laplace transform for a one unit step input is $1/s$. Hence

$$\text{Laplace transform of output} = G \times \frac{1}{\tau s + 1} \times \frac{1}{s}$$

$$= G \frac{(1/\tau)}{s[s + (1/\tau)]}$$

The transform is of the form

$$\frac{a}{s(s + a)}$$

where $a = (1/\tau)$. Hence

$$\theta_o = G[1 - e^{-t/\tau}]$$

Laplace transform for a second order system

The relationship between the input and output for a second order element is described by the differential equation

$$a_2 \frac{d^2\theta_o}{dt^2} + a_1 \frac{d\theta_o}{dt} + a_0\theta_o = b_0\theta_i$$

The Laplace transform if $\theta_o = 0$ and $d\theta_o/dt = 0$ at $t = 0$ is

$$a_2 s^2 \times \theta_o(s) + a_1 s \times \theta_o(s) + a_0 \times \theta_o(s) = b_0 \times \theta_i(s)$$

Hence

$$G(s) = \frac{\theta_o}{\theta_i} = \frac{b_0}{a_2 s^2 + a_1 s + a_0}$$

This can be rearranged to give

$$G(s) = \frac{(b_0/a_0)}{(a_2/a_0)s^2 + (a_1/a_0)s + 1}$$

As indicated earlier, the second order differential equation can be written in terms of the natural frequency ω_n and the damping ratio ζ. Then, as (b_0/a_0) is the static transfer function G,

$$G(s) = \frac{G}{(1/\omega_n^2)s^2 + (2\zeta/\omega_n)s + 1}$$

Consider the output when the system is subject to a step input.

$$\theta_o(s) = G(s) \times \theta_i(s)$$

Since $\theta_i(s) = 1/s$, then

$$\theta_o(s) = \frac{G}{[(1/\omega_n^2)s^2 + (2\zeta/\omega_n)s + 1]s}$$

Using partial fractions this can be rearranged as

$$\theta_o(s) = G\left[\frac{1}{s} - \frac{s + \zeta\omega_n}{(s + \zeta\omega_n)^2 + \omega_n^2(1 - \zeta^2)} - \frac{\zeta\omega_n}{(s + \zeta\omega_n)^2 + \omega_n^2(1 - \zeta^2)}\right]$$

When there is critical damping, i.e. $\zeta = 1$,

$$\theta_o(s) = G\left[\frac{1}{s} - \frac{1}{s + \omega_n} - \frac{\omega_n}{(s + \omega_n)^2}\right]$$

$$\theta_o = G[1 - e^{-\omega_n t} - \omega_n t e^{-\omega_n t}]$$
$$= G[1 - e^{-\omega_n t}(1 + \omega_n t)]$$

Transfer function for a system

As indicated in Chapter 1 the static transfer function G for a system is the product of the individual transfer functions of the elements of that system.

$$G = G_1 \times G_2 \times G_3$$

With the Laplace transform the system transfer function $G(s)$ is the product of the individual transfer functions

$$G(s) = G_1(s) \times G_2(s) \times G_3(s)$$

Thus, for example, a measurement system might consist of elements with the following transfer functions:

transducer: $\dfrac{G_1}{\tau_1 + 1}$

signal converter: $\dfrac{G_2}{\tau_2 s + 1}$

display: $\dfrac{G_3}{(\tau_3 + 1)^2}$

The system transfer function will thus be:

$$G(s) = \frac{G_1}{\tau_1 s + 1} \times \frac{G_2}{\tau_2 s + 1} \times \frac{G_3}{(\tau_3 + 1)^2}$$

5 Loading effects

Loading

When a thermometer at room temperature is put into hot water to measure its temperature, the act of putting the thermometer into the water changes the temperature of the water. Thus the act of attempting to make the measurement has modified the temperature being measured. Connecting an ammeter into an electrical circuit to measure the current changes the resistance of the circuit and so changes the current. The act of attempting to make the measurement has modified the current being measured. Such acts are known as *loading*.

Loading can also occur within a measurement system. This is when the connection of one element to the system modifies the characteristics of the preceding element.

Electrical loading

Thévenin's theorem can be stated as: an active network having two terminals A and B to which an electrical load may be connected (Figure 5.1), behaves as if the network contained a single source of e.m.f. E_{Th} in series with a single impedance Z_{Th}, where E_{Th} is the potential difference measured between A and B with the load disconnected and Z_{Th} the impedance of the network between A and B when all the sources of e.m.f. within the network have been replaced by their internal impedances.

Connecting a load Z_L across the output terminals of an active network is thus equivalent to connecting Z_L across the equivalent Thévenin circuit, as in Figure 5.1(b). The current i through Z_L is thus

$$i = \frac{E_{Th}}{Z_{Th} + Z_L}$$

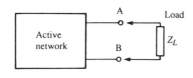

(a)

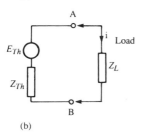

(b)

Figure 5.1 (a) Electrical circuit connected to load (b) Thévenin equivalent circuit

Hence the potential difference across the load V_L is given by

$$V_L = iZ_L = E_{Th} \frac{Z_L}{Z_{Th} + Z_L}$$

The effect of connecting the load across the network is thus to change the potential difference from E_{Th} to V_L. The value of V_L will approach that of E_{Th} the more Z_L is made greater than Z_{Th}. The condition for maximum voltage transfer is thus $Z_L \gg Z_{Th}$. Incidentally, the condition for maximum power transfer is that $Z_L = Z_{Th}$.

The effect of connecting a load across the network is thus to produce a loading error.

$$\text{Loading error} = E_{Th} - V_L$$

$$= E_{Th}\left(1 - \frac{Z_L}{Z_{Th} + Z_L}\right)$$

Loading of a voltmeter

When a voltmeter of resistance R_m is connected across a circuit with a Thévenin equivalent resistance of R_{Th} then the reading indicated by the instrument V_m is

$$V_m = E_{Th}\left(\frac{R_m}{R_m + R_{Th}}\right)$$

where E_{Th} is the Thévenin equivalent voltage of the circuit, i.e. the voltage before the meter was connected. Thus the accuracy of the voltmeter is

$$\text{accuracy} = \frac{V_m}{E_{Th}} \times 100\% = \frac{R_m}{R_m + R_{Th}} \times 100\%$$

Loading of a potentiometer

Figure 5.2(a) shows a potentiometer. The potentiometer slider is a distance x from one end of the potentiometer track, total track length being L. If the track has a uniform resistance per unit length, then open circuit voltage between terminals A and B is $(x/L)V_s$. This is then the Thévenin voltage.

$$E_{Th} = (x/L)V_s$$

The Thévenin impedance is found by setting V_s to be zero and calculating the impedance between terminals A and B (Figure 5.2(b)). This consists of two resistors in parallel, hence

$$\frac{1}{R_{Th}} = \frac{1}{(x/L)R_p} + \frac{1}{([L-x]/L)R_p}$$

where R_p is the total track resistance. Hence

$$R_{Th} = R_p(x/L)(1 - \{x/L\})$$

Figure 5.2(c) shows the Thévenin equivalent circuit. The current I in the circuit is given by

$$(x/L)V_s = I[R_p(x/L)(1 - \{x/L\}) + R_L]$$

Hence the potential difference across the load is

$$V_L = IR_L = \frac{R_L(x/L)V_s}{R_p(x/L)(1 - \{x/L\}) + R_L}$$

$$= \frac{(x/L)V_s}{(R_p/R_L)(x/L)(1 - \{x/L\}) + 1}$$

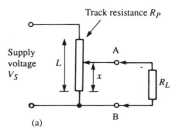

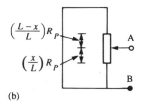

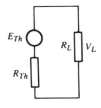

Figure 5.2 Potentiometer: (a) the circuit (b) R_{Th} (c) the Thévenin circuit

The relationship between V_L and x is non-linear. Hence the effect of the loading is to give a non-linearity error.

$$\text{Non-linearity error} = E_{Th} - V_L$$

$$= (x/L)V_s\left[1 - \frac{1}{(R_p/R_L)(x/L)(1 - \{x/L\}) + 1}\right]$$

If $R_L \gg R_p$ the equation approximates to

$$\text{Non-linearity error} = V_s(R_p/R_L)(\{x/L\}^2 - \{x/L\}^3)$$

The maximum value of this error occurs when $d(\text{error})/dx = 0$, i.e. when $(X/L) = 2/3$. Hence

$$\text{maximum non-linearity error} = 0.148 V_s(R_p/R_L)$$

Loading of a Wheatstone bridge

The Thévenin voltage for a Wheatstone bridge (Figure 5.3(a)) is the open circuit voltage when $I = 0$. Hence, since V_s is the potential difference between A and C,

$$V_s = I_1(R_1 + R_2)$$
$$V_s = I_2(R_3 + R_4)$$

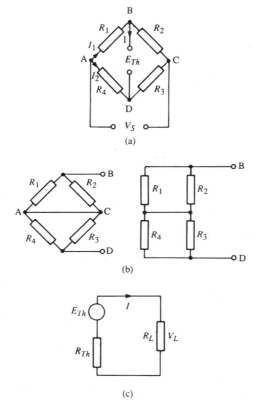

Figure 5.3 Wheatstone bridge: (a) E_{Th} (b) R_{Th} (c) Thévenin circuit with load

The potential difference between A and B, i.e. across R_1, is $I_1 R_1$ and that between A and D, i.e. across R_4 is $I_2 R_4$. Thus the potential difference between B and D is

$$E_{Th} = I_1 R_1 - I_2 R_4$$
$$= \frac{V_s R_1}{R_1 + R_2} - \frac{V_s R_4}{R_3 + R_4}$$

The Thévenin resistance R_{Th} between B and D (Figure 5.3(b)) is that of a pair of parallel resistors R_1 and R_2 in series with the parallel resistor pair R_3 and R_4. Thus

$$R_{Th} = \frac{R_1 R_2}{R_1 + R_2} + \frac{R_3 R_4}{R_3 + R_4}$$

E_{Th} is the output potential difference when there is no load. With a load resistance the circuit is as shown in Figure 5.3(c). Then

$$E_{Th} = I(R_L + R_{Th})$$

The potential difference across the load V_L is then the output.

$$V_L = IR_L = \frac{E_{Th}R_L}{R_L + R_{Th}}$$

$$= \frac{R_L V_s[R_1/(R_1 + R_2) - R_4/(R_3 + R_4)]}{R_L + [R_1 R_2/(R_1 + R_2) + R_3 R_4/(R_3 + R_4)]}$$

$$= \frac{R_L R_s(R_1 R_3 - R_2 R_4)}{R_L(R_1 + R_2)(R_3 + R_4) + R_1 R_2 + R_3 R_4}$$

Loading of elements in a measurement system

Consider a measurement system (Figure 5.4) consisting of a transducer, an amplifier and an indicator. The transducer has an open-circuit output voltage of V_t and a resistance R_t. The amplifier has an input resistance of R_{in}. This is the load across the transducer. Hence the potential difference V_{in} across this load is

$$V_{in} = I_1 R_{in} = \frac{V_t R_{in}}{R_t + R_{in}}$$

If the amplifier has a transfer function of G then GV_{in} will be the open-circuit output from the amplifier. The amplifier has an output resistance of R_{out}. The indicator is a load of resistance R_I. Hence the output potential difference from the indicator V_I is

$$V_I = I^2 R_I = \frac{GV_{in}R_I}{R_{out} + R_I}$$

$$= \frac{GV_t R_{in}R_I}{(R_t + R_{in})(R_{out} + R_I)}$$

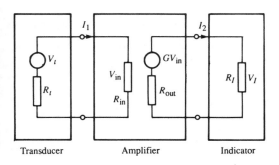

Transducer   Amplifier   Indicator

Figure 5.4 Measurement system loading

6 Noise

Noise

The term noise is generally used for the unwanted signals that may be picked up by the measurement system and interfere with the signal being measured. There are two types of noise:

1 *Interference.* This is due to the interaction between external electrical and magnetic fields and the measurement system circuits, e.g. the circuit picking up interference from nearby mains power circuits.
2 *Random noise.* This noise is due to the random motion of electrons and other charge carriers in components and is a characteristic of the basic physical properties of components in the system.

Types of interference

The three main types of interference are:

1 *Inductive coupling.* This is sometimes referred to as *electromagnetic coupling* or *magnetic coupling*. A changing current in a nearby circuit produces a changing magnetic field. The effect of the changing magnetic field on conductors within the measurement system is to induce e.m.f.s in them.
2 *Capacitive coupling.* Nearby power cables, the earth, and conductors in the measurement system are separated by a dielectric, air. Thus there can be capacitance between the power cable and conductors, and between the conductors and earth. These capacitors couple the measurement system conductors to the other systems and thus signals in those systems pass to the measurement system as interference.
3 *Multiple earths.* If the measurement system has more than one connection to earth there may be problems since there may be some difference in potential between the earth points. If this occurs the earthing may produce an interference current through the measurement system.

Reduction of interference

Methods of reducing interference are:

1 *Twisted pairs of wires.* The elements of the measurement system are connected by twisted wire pairs (Figure 6.1). A changing magnetic field will induce e.m.f.s in each loop but because of the twisting the directions of the e.m.f.s in a wire will be for the portion in one loop in one direction and then in the next loop in the opposite direction. The result is cancellation of the induced e.m.f.s.
2 *Electrostatic screening.* The ideal method of avoiding

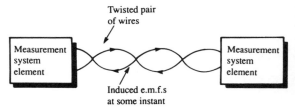

Figure 6.1 Reduction of interference by twisted pairs

capacitance coupling is to completely enclose the transducer and the entire measurement system in an earthed metal screen. There may be problems of multiple earths if, for instance, the transducer is separately earthed from the earthed display. Coaxial cable gives screening of connections between elements of measurement systems, however the cable should only be earthed at one end if multiple earths are to be avoided.

3 *Single earth.*   Multiple earthing can be avoided by having only one earthing point.

4 *Differential amplifiers.*   A differential amplifier can be used to amplify the difference between two signals. Thus if both signals contain the same interference then the output from the amplifier will not have amplified any interference signals.

5 *Filters.*   A filter can be selected which transmits the measurement signal but rejects the interference signal.

Further reading: Putten, A. F. P. van, (1988), *Electronic Measurement Systems*, Prentice Hall.

Cross talk

In some measurement systems inputs from a number of transducers may be fed into the system through a multiconductor cable or a ribbon conductor. The term *cross talk* is used to describe the interference that occurs between these signals. The interference is a mixture of inductive and capacitive coupling. The interference may be reduced by increasing the spacing between the conductors, screening the most affected circuits, or with a ribbon conductor, interspersing a signal carrying conductors with earthed conductors.

Random noise

Random noise can arise in a number of ways.

1 *Thermal noise*, sometimes referred to as *Johnson noise.*   This is noise generated by the random motion of electrons and other charge carriers in resistors and semiconductors. It is spread over an infinite range of frequencies and is thus referred to as *white noise*. The r.m.s. noise voltage for a bandwidth of frequency f_1 to f_2 is

$$\sqrt{[4kRT(f_2 - f_1)]}$$

where k is Boltzmann's constant, R the resistance and T the absolute temperature. Thus a wideband amplifier will generate more noise than a narrowband one. High resistances and high temperatures will also result in more noise.

2 *Shot noise.*   This is noise due to the random fluctuations in the rate at which charge carriers diffuse across potential barriers such as in a p–n junction. The r.m.s. noise voltage is, for an absolute temperature T and bandwidth of frequency f_1 to f_2,

$$\sqrt{[2kTr_d(f_2 - f_1)]}$$

where k is Boltzmann's constant and r_d the differential diode resistance, this being kT/qI where q is the charge on the electron and I the d.c. current in the junction.

3 *Flicker noise.*   This is noise due to a flow of charge carriers in a discontinuous medium, e.g. a carbon composite resistor. The r.m.s. noise voltage is approximately inversely proportional to the frequency.

4 *Poor connections.*   Noise can result from poor connections due to dirt on switch contacts or bad soldering.

Further reading: Putten, A. F. P. van, (1988), *Electronic Measurement Systems*, Prentice Hall.

Noise rejection

The term *normal mode noise* is used to describe all the noise occurring within the signal source. To the measurement system this noise is indistinguishable from the actual quantity that it is set up to measure. The ability of a system to reject normal mode noise is called the *normal mode rejection ratio* (N.M.R.R.) or *series mode rejection ratio*. This can be defined, in decibels, as

$$\text{N.M.R.R.} = 20 \log_{10}\left(\frac{V_n}{V_e}\right)$$

where V_n is the peak value of the normal mode noise and V_e is the peak value of the error it produces in the measurement at a particular frequency. An alternative way of describing the N.M.R.R. is in terms of the peak value of the normal mode noise which would not produce an error greater than some specified error value.

The term *common mode noise* is used to describe the noise occurring between the earth terminal of a measurement system and its lower potential terminal. The ability of a measurement system to prevent common mode noise introducing an error in the measurement reading is called the *common mode rejection ratio* (C.M.R.R.).

$$\text{C.M.R.R.} = 20 \log_{10}\left(\frac{V_{cm}}{V_e}\right)$$

where V_{cm} is the peak value of the common mode noise and V_e the peak value of the error it produces in the measurement at a particular frequency.

Signal-to-noise ratio

The signal-to-noise ratio is the ratio of the signal power to the noise power.

$$\text{S/N ratio} = \frac{\text{signal power}}{\text{noise power}}$$

This is usually expressed in decibels, hence

$$\text{S/N ratio} = 10 \log_{10}\left(\frac{\text{signal power}}{\text{noise power}}\right)$$

Since the power is V^2/R then if V_s is the signal voltage and V_n the noise voltage

$$\text{S/N ratio} = 10 \log_{10}\left(\frac{V_s}{V_n}\right)^2$$

$$\text{S/N ratio} = 20 \log_{10}\left(\frac{V_s}{V_n}\right)$$

Noise factor

The noise factor or *noise figure* F is, when expressed in dB,

$$F = 10 \log_{10} (\text{total noise power output/input noise})$$

The noise factor is thus a measure of the amount of noise introduced by the element. If the input noise power is P_{ni} and the element has a transfer function G then the output noise power due to the input noise power is GP_{ni}. But G is the output signal power P_{so} divided by the input signal power P_{si}. Hence

$$F = 10 \log_{10}\left[\frac{P_{no}}{(P_{so}/P_{si})P_{ni}}\right]$$

$$= 10 \log_{10} \left[\frac{P_{si}/P_{ni}}{P_{so}/P_{no}} \right]$$

$$= \frac{\text{input signal-to-noise ratio}}{\text{output signal-to-noise ratio}}$$

when the signal-to-noise ratios are expressed in dB.

Further reading: Putten, A. F. P. van, (1988), *Electronic Measurement Systems*, Prentice Hall.

Averaging

Averaging can be used to enhance the signal-to-noise ratio for a repetitive signal. For the same point in the signal waveform samples are taken for a number of cycles and the average value obtained. Because of the random nature of the noise signal superimposed on the measurement signal the noise component in each sample will be different, sometimes negative sometimes positive. The result of the averaging is thus to give a reduced average noise value. This averaging process is carried out for a number of points in the waveform and the signal reconstructed. The improvement in signal-to-noise ratio is proportional to the square root of the number of points of the measurement signal sampled.

7   Reliability

Reliability and unreliability

The *reliability* of a measurement system or measurement system element can be defined as the chance that the system or element will operate to a specified level of performance for a specified period under specified environmental conditions. The specified level of performance might be an accuracy of $\pm 1\%$, performance outside this would be considered to be a failure. The specified environmental conditions might be at a temperature of $20\,°C$.

The *unreliability* of a measurement system or measurement system element is the chance that the system or element will fail to operate to the specified level of performance for a specified period under specified environmental conditions.

Chance, or probability, is the frequency in the long run with which an event occurs. In the case of reliability this is survival, in the case of unreliability it is failure. Thus, if we start off with N_0 items and after some time t there are N left,

$$\text{reliability} = \frac{N}{N_0}$$

The number failing in that time is $(N_0 - N)$. Hence

$$\text{unreliability} = \frac{N_0 - N}{N_0}$$

$$= 1 - \frac{N}{N_0}$$

$$= 1 - \text{reliability}$$

Further reading: Putten, A. F. P. van (1988), *Electronic Measurement Systems*, Prentice Hall; Smith, D. J. (1985), *Reliability and Maintainability*, Macmillan.

Failure

Failure is when a measurement system or measurement system element fails to perform to the specified level of performance. If N items are tested for a time t with failed items being repaired and put back into service, then if during that time there are N_f failures the *mean time between failures* (M.T.B.F.) is

$$\text{M.T.B.F.} = \frac{Nt}{N_f}$$

The *failure rate* λ is the average number of failures, per item, per unit time. Thus if N items are tested for a time t with failed items being repaired and put back into service, then if during that time there are N_f failures the failure rate is

$$\lambda = \frac{N_f}{Nt}$$

Hence

$$\lambda = \frac{1}{\text{M.T.B.F.}}$$

Table 7.1 shows some typical values of failure rates for components.

Table 7.1 Failure rates

Component	Failure rate $\times 10^{-5}$ per hour
Carbon resistor	0.05
Wire wound resistor	0.01
Paper capacitor	0.1
Plastic film capacitor	0.01
Silicon transistor ($>1\,$W)	0.08
Silicon transistor ($<1\,$W)	0.008
Filament lamp	0.5
Soldered connection	0.001
Wrapped connection	0.0001

A failure rate of, say, $1/100\,000$ per hour does not mean that if $100\,000$ items are observed for one hour that exactly one will fail. The failure rate is the average value for large numbers of items and large numbers of failures. Thus the failure rate obtained from a test with a number of items is, because of the finite size of the sample, only an estimate of the likely failure rate. So that the failure rate is not underestimated the observed failure rate is generally multiplied by some factor to give the assessed failure rate. The factor used depends on the *confidence level* required. A confidence level of, say, 60% means that in at least 60% of the cases considered the true failure rate will not exceed the assessed failure rate.

Availability

The availability of a system is the probability that the system is in an operational state at any specific time. This is the fraction of the time for which the system is in an operational state. Thus

$$\text{availability} = \frac{\text{time operational}}{\text{time operational} + \text{time not operational}}$$

$$= \frac{\text{MTBF}}{\text{MTBF} + \text{MTTR}}$$

where MTBF is the mean time between failures and MTTR is the mean time to repair.

Failure rate and time

Figure 7.1 shows how the failure rate for a given type of measurement system or measurement system element commonly varies with time. Because of its shape the graph is referred to as the 'bath tub' graph. There are three distinct phases:

1 Early failure, or the burn-in period, due to manufacturing faults and material imperfections.
2 Normal working life, where the failure rate is virtually constant and the result of purely random events.
3 Wear out failure, where the failure rate increases due to components wearing out.

The exponential law of reliability

Consider there to be a constant failure rate λ. If at time $t = 0$ there are N_0 items then by time t the number of items will have decreased to N, when no items are being repaired. After a further interval of time δt the

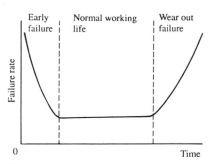

Figure 7.1 The 'bath tub' graph

number will have dropped to $(N - \delta N)$. The number failing is thus $-\delta N$. Thus the failure rate is

$$\lambda = \frac{-\delta N}{N \delta t}$$

Thus in the limit when $\delta t \to 0$,

$$\lambda = -\frac{1}{N} \frac{dN}{dt}$$

$$\frac{dN}{N} = -\lambda \, dt$$

Integrating gives

$$\int_{N_0}^{N} \frac{dN}{N} = -\int_{0}^{t} \lambda \, dt$$

$$\ln N - \ln N_0 = -\lambda t$$
$$N = N_0 e^{-\lambda t}$$

The reliability is N/N_0, hence

$$\text{reliability} = e^{-\lambda t}$$

The total number of failures in the time t is $(N_0 - N)$. Hence

$$\text{number of failures} = N_0 - N_0 e^{-\lambda t}$$

The unreliability is (number of failures/N_0). Hence

$$\text{unreliability} = 1 - e^{-\lambda t}$$

Figure 7.2 shows graphs of the reliability and unreliability variations with time.

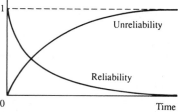

Figure 7.2 Variation of reliability and unreliability with time

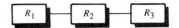

Figure 7.3 A system of three elements in series

Reliability of a system

Consider a system consisting of a number of elements in series, as in Figure 7.3. Each of the elements has its own, independent, reliability. The entire system will, however, fail if one of the elements fails. Hence the reliability R of a three element system is

$$R = R_1 \times R_2 \times R_3$$

where R_1, R_2 and R_3 are the reliabilities of the elements.

The reliabilities are multiplied because the probability of three things happening together is the product of their separate probabilities. For example, the chance of one coin landing with heads uppermost is 1/2. The chance of three coins landing with heads uppermost is $(1/2) \times (1/2) \times (1/2)$ or 1/8. This can be worked out by considering all the possible ways the coins can land. Only one way in eight is all heads.

```
H T T     H H T     H H H     H T H
T T T     T H T     T H H     T T H
```

The reliability of the system R_s can be expressed in terms of the system failure rate λ_s as

$$R_s = \exp(-\lambda_s t)$$

Hence since $R_1 = \exp(-\lambda_1 t)$, $R_2 = \exp(-\lambda_2 t)$ and $R_3 = \exp(-\lambda_3 t)$ then

$$\begin{aligned} \exp(-\lambda_s) &= \exp(-\lambda_1 t) \times \exp(-\lambda_2 t) \times \exp(-\lambda_3 t) \\ &= \exp-(\lambda_1 + \lambda_2 + \lambda_3)t \end{aligned}$$

Hence

$$\lambda_s = \lambda_1 + \lambda_2 + \lambda_3$$

The failure rate of the system is the sum of the failure rate of the constituent elements.

Redundancy

The reliability of a system can be improved by using more reliable elements. Another way is to introduce redundancy in the system. This means for a particular element that an alternative element is in parallel with it (Figure 7.4). The system can only then fail if both elements fail.

The unreliability of one of the parallel elements is $(1 - R_1)$, the unreliability of the other $(1 - R_2)$. Unreliability is the chance of failure. The probability that both elements will fail is the product of the individual probabilities because the probability of two independent

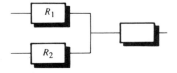

Figure 7.4 Redundancy

events occurring is the product of the individual probabilities. Hence the unreliability of both elements failing is

unreliability of system $= (1 - R_1)(1 - R_2)$

Hence the reliability of the arrangement is

reliability $= 1 - (1 - R_1)(1 - R_2)$
$= R_1 + R_2 - R_1 R_2$

Part Two

System Components

8 Transducers

Table 8.1 indicates the forms of the input and output signals for the transducers referred to in this chapter.

Table 8.1 Transducers

Form of variable		Transducer
Input	Output	
Acceleration	Charge	15
Alternating p.d.	Ultrasonic wave	16
Angular displacement	Angular displacement	13
Angular displacement	Potential difference	4
Angular displacement	Capacitance	7
Angular displacement	Pulses	28, 29
Angular position	Pulses	29
Angular velocity	Pulses/alternating e.m.f.	10
Density	Frequency	31
Displacement	Capacitance	6
Displacement	Capacitance	8
Displacement	Inductance	9
Displacement	Inductance difference	11
Displacement	Potential difference	12
Displacement	Pressure	25
Fluid flow rate	Pulses/alternating e.m.f.	27
Fluid flow rate	Pressure	26
Fluid velocity	Resistance	2
Force	Charge	15
Force	Displacement	20, 21
Force	Frequency	30
Force	Resistance	20, 21
Light intensity	Current	18
Light intensity	Potential difference	17
Light intensity	Resistance	5
Liquid level	Capacitance	8
pH	Potential difference	19
Pressure	Capacitance	6
Pressure	Charge	15
Pressure	Displacement	22, 23, 24
Strain	Resistance	3
Temperature	E.m.f.	14
Temperature	Resistance	1
Vibration	Charge	15

Resistive transducers

1 Temperature sensors

The relationship between the resistance of a metal and temperature is of the form

$$R_t = R_0(1 + \alpha t + \beta t^2 + \gamma t^3 + \ldots)$$

where R_t is the resistance of a length of wire at temperature $t\,°C$. R_0 its resistance at $0\,°C$ and α, β and γ are temperature coefficients of resistance, with $\alpha > \beta > \gamma$. For most metals the resistance increases at a reasonably linear rate with temperature and β, γ and higher terms can be neglected. For such a linear relationship

$$R_t = R_0(1 + \alpha t)$$

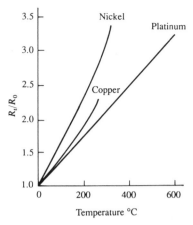

Figure 8.1 Resistance–temperature graphs for platinum, nickel and copper

Table 8.2   Temperature coefficients of resistance

Metal	$\alpha \,^\circ C^{-1}$
Copper	3.8×10^{-3}
Nickel	6.7×10^{-3}
Platinum	3.9×10^{-3}

Figure 8.1 shows the resistance–temperature graphs, and Table 8.2 the temperature coefficients of resistance, for three metals commonly used as transducers.

The resistance of semiconductors also changes with temperature. A group of transducers based on this are *thermistors*. These are made from mixtures of metal oxides, such as those of chromium, cobalt, iron, manganese and nickel formed into various forms such as beads, discs and rods. The resistance–temperature graph for a thermistor is highly non-linear and is described by the exponential relationship

$$R_t = K e^{\beta/t}$$

where R_t is the resistance at temperature t, with K and β being constants. The resistance of thermistors generally decreases with an increase in temperature (Figure 8.2) though there are some for which the resistance increases with an increase in temperature. Whether increasing or decreasing, the change in resistance per degree change in temperature is considerably larger than that which occurs with metals. For thermistors which show a decrease in resistance with an increase in temperature self heating due to currents can become significant. This means that the current may be large enough for the power developed to raise the temperature of the thermistor above that of its environment. This increase in temperature results in a decrease in resistance and consequently the current increases yet further and so the temperature is increased even more. This effect continues until the heat dissipation of the thermistor equals the power supplied to it.

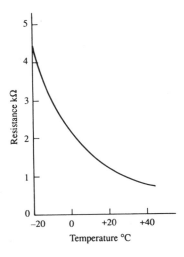

Figure 8.2 Resistance–temperature graph for a thermistor

2 Hot wire anemometer

A current i through a resistance R generates heat of i^2R. When the resistive element is immersed in a fluid the heat lost by it is $hA(\theta_t - \theta_f)$, where h is the heat transfer coefficient, A the effective surface area of the resistor, θ_t the temperature of the resistive element and θ_f that of the fluid. Equilibrium occurs when

$$i^2R = hA(\theta_t - \theta_f)$$

The heat transfer coefficient depends on the fluid velocity v,

$$h = a + b\sqrt{v}$$

where a and b are constants for a particular fluid. Thus

$$i^2R = (a + b\sqrt{v})A(\theta_t - \theta_f)$$

One way of using the resistive element as a transducer to determine fluid velocities is to maintain its temperature θ_t constant. This is done by using a network which maintains the resistance R constant, since its resistance depends on its temperature. Then

$$i^2 = A + B\sqrt{v}$$

where A and B are constants. Such a transducer is known as a *constant temperature anemometer*.

3 Strain gauges

When a length of wire, or metal foil or semiconductor, is stretched its resistance changes. The fractional change in resistance, $\Delta R/R$, is directly proportional to the strain ε

$$\Delta R/R = G\varepsilon$$

where G is a constant called the gauge factor. For most materials the gauge factor is positive, indicating that the resistance increases when the strain increases, i.e. tension increases resistance, compression decreases it.

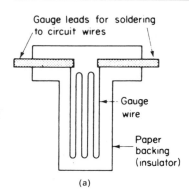

Gauge leads for soldering to circuit wires

Gauge wire

Paper backing (insulator)

(a)

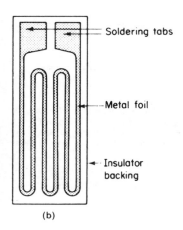

Soldering tabs

Metal foil

Insulator backing

(b)

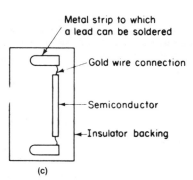

Metal strip to which a lead can be soldered

Gold wire connection

Semiconductor

Insulator backing

(c)

Figure 8.3 Strain gauges (a) metal wire (b) metal foil (c) semiconductor

The term *strain gauge* is used for the element which can be stuck onto surfaces like a postage stamp and whose resistance changes when subject to strain. The wire strain gauge consists of a length of wire wound in a grid shape and attached to a suitable backing material (Figure 8.3(a)) and has a gauge factor of about 2.0. The metal foil strain gauge consists of a grid form which has been etched from a metal foil (Figure 8.3(b)) and mounted on a resin film base. It has a gauge factor of about 2.0. Semiconductor strain gauges (Figure 8.3(c)) are strips of silicon doped with small amounts of p or n-type material. The p-type silicon gives gauge factors of about 100 to 175, the n-type silicon -100 to -140. The negative gauge factor means that the resistance decreases with an increase in strain. Semiconductors have the advantage over metal gauges of a high gauge factor but have the disadvantage of a much greater temperature coefficient of resistance.

The resistance of a strain gauge is changed not only by a change in strain but also a change in temperature. See Chapter 9 for details of ways of compensating for the effects of temperature.

4 Potentiometers

For a constant current through a resistor the potential difference across it depends on the value of the resistance. This principle is used with the *potentiometer*. The potentiometer consists of a circular wire wound track or a film of conductive plastic over which a rotatable electrical contact, the slider, can be rotated and so vary the resistance, and hence the potential difference, between the output terminals 1 and 2 (Figure 8.4). The track may be just a single turn or helical.

If the track has a constant resistance per unit length then the output V_o is proportional to the length of track between terminals 1 and 2 and hence, with a circular potentiometer track, is proportional to the angle θ through which the slider has been rotated.

$$V_o = k\theta$$

where k is a constant. Hence an angular displacement can be converted into a potential difference.

The following are some typical characteristics of potentiometers: resolution with a wire track about 1.5 mm for a coarsely wound track to 0.5 mm for a finely wound one, conductive plastic track infinite resolution; non-linearity errors from less than 0.1% to about 1% for wire tracks, about 0.05% for conductive plastics; track resistance from about 20 Ω to 200 kΩ for wire tracks, from about 500 Ω to 80 kΩ for conductive plastics. The conductive plastic has a higher temperature coefficient of resistance than the wire and so temperature changes have a greater effect on the accuracy. The characteristics of a potentiometer are affected by the resistance of the load connected across the output, see Chapter 5.

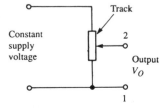

Figure 8.4 Potentiometer

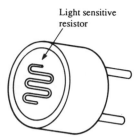

Figure 8.5 Photoconductive cell

Table 8.3 Photoconductive cells

Photoconductor	Spectral range (μm)
Cadmium sulphide	0.47 to 0.71
Cadmium selenide	0.60 to 0.77
Lead sulphide	1.0 to 3.0
Lead selenide	1.5 to 4.0

Note: the visible spectrum extends from 0.40 to 0.76 μm.

5 Photoconductive cells

A *photoconductive cell* (Figure 8.5) is a light-dependent resistor. Its resistance decreases as the intensity of light falling on it increases. Cadmium sulphide is a commonly used cell because it has a response to the colours of the spectrum which is very similar to that of the human eye (see Table 8.3), being most sensitive at 0.6 μm. Lead sulphide cells can be used out into the infrared. The resistance of a cell typically varies from megohms in the dark to a few hundred ohms in bright light. The response time to a change in light is typically of the order of 50 ms.

Capacitive transducers

A capacitor formed by two parallel plates separated by a dielectric has a capacitance C given by

$$C = \frac{\varepsilon_r \varepsilon_0 A}{d}$$

where ε_r is the relative permittivity of the dielectric between the capacitor plates, ε_0 the permittivity of free space (8.85×10^{-12} F/m), A the area of overlap between the two plates and d the plate separation. Changes in the plate separation, the area of overlap between plates and the dielectric can all be used as the basis of transducers.

6 Variable plate separation transducer

A change in the separation of capacitor plates produces a change in capacitance, thus changes in the position of one plate when the other is fixed give rise to changes in capacitance. A pressure gauge based on this consists of a circular diaphragm acting as one plate of the capacitor and a fixed plate for the other one (Figure 8.6). When the pressure changes the diaphragm distorts, changing the separation between it and the fixed plate and hence the capacitance.

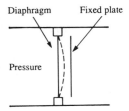

Figure 8.6 A variable plate separation pressure transducer

Item 6 in Chapter 9, Figure 9.10, gives another form of variable plate separation transducer, a push–pull form.

7 *Variable plate area transducer*

A change in the area of overlap of a pair of capacitor plates will change the capacitance. Figure 8.7 shows how the principle can be used to convert an angular displacement into a change in capacitance.

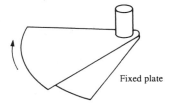

Figure 8.7 Variable plate area capacitive transducer

8 *Variable dielectric transducer*

If the relative amount of two dielectrics between capacitor plates varies, in the way shown in Figure 8.8, then the capacitance varies. The arrangement can be considered to be effectively two capacitors in parallel. Hence the total capacitance is the sum of the capacitance of the two capacitors formed by the two dielectrics. If w is the width of the capacitor plates and the dielectrics, then

$$C = \frac{\varepsilon_1 \varepsilon_0 w x}{d} + \frac{\varepsilon_2 \varepsilon_0 w (L - x)}{d}$$

$$= \frac{\varepsilon_0 w}{d} [\varepsilon_2 L - (\varepsilon_2 - \varepsilon_1) x]$$

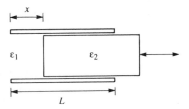

Figure 8.8 Variable dielectric capacitive transducer

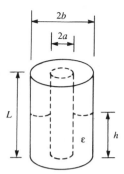

Figure 8.9 Capacitive liquid level gauge

The capacitance thus depends on the displacement x and can be used for its measurement.

Figure 8.9 shows a version of this transducer for the measurement of liquid level, i.e. displacement of the liquid surface. The plates of the capacitor are two concentric conducting cylinders with the two dielectrics between them being the liquid and the air above the liquid surface. The capacitance per unit length C of coaxial cylinders, radii a and b, is

$$C = \frac{2\pi\varepsilon_r\varepsilon_0}{\ln(b/a)}$$

where ε_r is the relative permittivity of the medium between the cylinders. The total capacitance is the sum of the capacitances of the two capacitors, one being of length h and the other $(L-h)$.

$$C = \frac{2\varepsilon_r\varepsilon_0 h}{\ln(b/a)} + \frac{2\varepsilon_0(L-h)}{\ln(b/a)}$$

$$C = \frac{2\pi\varepsilon_0}{\ln(b/a)}[L + (\varepsilon_r - 1)h]$$

Inductive transducers

9 Variable reluctance
The flux in a series magnetic circuit depends on the reluctances of the elements in the circuit. For the circuit shown in Figure 8.10 part of the circuit is the air between the ferromagnetic plate and the ferromagnetic core. The length of this air path is changed by displacement of the plate. Hence the flux in the circuit changes. A change in the flux linked by the coil results in a change in its inductance. Hence the inductance becomes a measure of the displacement of the plate.

The total reluctance S of the series circuit is the sum of the reluctances of the core, plate and air gaps. When the air gap is zero, i.e. $d = 0$, if the reluctance of the core plus plate is S_0 then the inductance of the coil is

$$L_0 = \frac{N^2}{S_0}$$

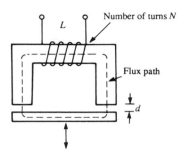

Figure 8.10 Variable reluctance transducer

When there is an air gap then the reluctance of the air element in the magnetic circuit is

$$S_a = \frac{2d}{\mu_0 A}$$

where μ_0 is the permeability of free space ($4\pi \times 10^{-7}$ H/m), $2d$ the length of the magnetic flux path in the circuit and A the cross-sectional area of the flux path. The total reluctance now of the circuit is thus

$$S = S_0 + S_a = S_0 + \frac{2d}{\mu_0 A}$$

The inductance of the coil becomes

$$L = \frac{N^2}{S_0 + (2d/\mu_0 A)} = \frac{L_0}{1 + (2d/\mu_0 A S_0)}$$
$$= \frac{L_0}{1 + kd}$$

where k is a constant. The relationship between L and d is non-linear.

Item 6 in Chapter 9, Figure 9.11, gives another form of this transducer, a push-pull form.

10   Variable reluctance tachogenerator

Another version of the variable reluctance transducer, a variable reluctance tachogenerator, is used for the measurement of the angular speed of a rotating shaft. It consists of a toothed ferromagnetic wheel which rotates with the shaft, and a detector, a permanent magnet around which a coil is wound, which produces a voltage pulse each time a tooth passes it (Figure 8.11). The arrangement is a magnetic circuit which has an air gap, the size of the gap depending on whether a tooth or a gap is opposite the magnet. The circuit reluctance changes each time a tooth passes the magnet. Hence the flux passing through the coil fluctuates about a mean value. This fluctuation can be considered to be almost sinusoidal. Such a flux change induces an alternating e.m.f. in the coil, both the frequency and the amplitude of this e.m.f. being proportional to the angular velocity of the wheel.

If the wheel contains n teeth and rotates with an angular velocity ω then the flux change with time for the coil can be considered to be of the form

$$\phi = \phi_0 + \phi_a \cos n\omega t$$

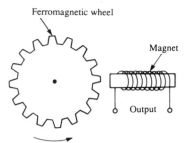

Figure 8.11 Variable reluctance tachogenerator

where ϕ_0 is the mean value of the flux and ϕ_a the amplitude of the flux variation. The induced e.m.f. E by the N turns of the coil is

$$E = -\frac{N\mathrm{d}\phi}{\mathrm{d}t} = -\frac{N\mathrm{d}}{\mathrm{d}t}(\phi_0 + \phi_a \cos n\omega t)$$

$$= N\phi_a n\omega \sin n\omega t$$

This is an equation of the form

$$E = E_{max} \sin 2\pi ft$$

The maximum value of this e.m.f. is thus $N\phi_a n\omega$ and its frequency f is $nw/2\pi$.

An alternative way of processing the output from the coil is to use a pulse shaping element to transform the output into pulses which can be counted by a counter.

11 Variable differential inductor

A variable differential inductor consists of two identical coils between which a core rod of high permeability material is moved (Figure 8.12).

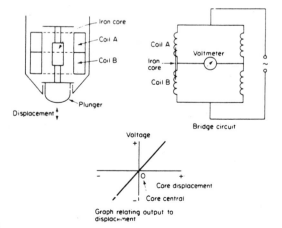

Figure 8.12 Variable differential inductor

The inductance of each coil depends on the length of the core rod inside it. When the core rod has the same length in each coil they have the same inductance. Any movement of the rod then results in the inductance of one coil increasing and the other decreasing. The difference in inductance between the coils is thus a measure of the displacement of the rod.

12 Variable differential transformer

The linear variable differential transformer (LVDT) is a transformer with a primary coil and two secondary coils (Figure 8.13). The two secondary windings are connected in series so that their outputs oppose each other. An alternating voltage input to the primary coil induces alternating e.m.f.s in the secondary coils. Both the secondary coils are identical so that when the core is central with equal amounts in each secondary coil the e.m.f.s induced in the secondary coils are the same. Since they are so connected that their outputs oppose each other, the result is an overall zero output. A change in the position of the core so that there are different amounts in the two secondary coils results in the induced e.m.f. in one coil becoming greater than that in the other and so a net output.

For a sinusoidal input voltage the e.m.f.s in the two secondary coils can be represented by

$$V_A = k_A \sin(\omega t - \phi)$$
$$V_B = k_B \sin(\omega t - \phi)$$

where the values of k_A and k_B depend on the degree of coupling occurring between the primary and secondary coils for a particular core position. ϕ is the phase difference between the primary alternating voltage and the secondary voltages. Thus the output V_o is

$$V_o = V_A - V_B = (k_A - k_B) \sin(\omega t - \phi)$$

When the core is at its mid position with equal amounts in each coil then k_A equals k_B and so V_o is zero. When the core is more in A than B then $k_A > k_B$, when more in B than A then $k_A < k_B$. A consequence of this is that there is a phase change of $180°$ in the output when the core moves from more in A to more in B.

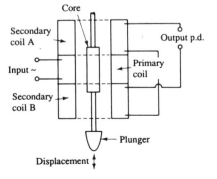

Figure 8.13 Linear variable differential transformer

For $k_A > k_B$

$$V_o = (k_A - k_B)\sin(\omega t - \phi)$$

For $k_A < k_B$

$$V_o = -(k_A - k_B)\sin(\omega t - \phi)$$
$$= (k_B - k_A)\sin[\omega t + (\pi - \phi)]$$

The a.c. characteristic is thus of the form shown in Figure 8.14(a). The output voltage can be converted into d.c. in a way which distinguishes between the phases and so produce a d.c. characteristic of the form shown in Figure 8.14(b). Both the characteristics shown in the figure are ideal ones since there is inevitably some non-linearity near both extremes of core displacement. The LVDT is used to measure displacements from about a quarter of a millimetre to 250 mm.

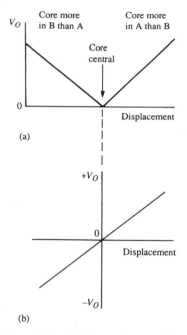

Figure 8.14 LVDT (a) a.c. characteristic (b) d.c. characteristic

13 Synchro

A synchro element, or a.c. position motor, consists of three stator windings at 120° spacing around the stator case (Figure 8.15). An a.c. input to the rotor coil induces outputs in each of the secondary coils. The relationship between the outputs from the three stator coils depends on the angular position of the rotor.

Synchro elements are frequently used in pairs (Figure 8.16). One acts as a transmitter and the other as a receiver. When the rotors of the two elements are in the same angular positions with reference to their

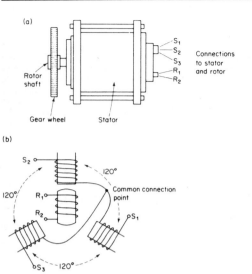

Figure 8.15 Synchro (a) external appearance (b) internal arrangement

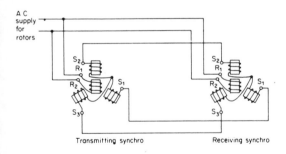

Figure 8.16 A synchronous transmission link

stator coils then the e.m.f.s in corresponding coils are identical and no potential differences exist between corresponding pairs. No current then flows between the coils. However, if the rotors are not in the same angular positions the e.m.f.s induced in corresponding pairs of coils are not the same. Currents then pass between coils and produce magnetic fields which in turn produce forces on the rotors which cause them to become aligned. Thus movement of the rotor of one synchro results in a movement of the other rotor to the same angular position. This means that the transmitting rotor can be mechanically coupled to a shaft and, some distance away, the receiving rotor used to move a pointer across a scale. The angular position of the shaft thus leads to a corresponding angular position of a pointer moving across a scale.

Thermoelectric transducers

14 Thermocouples

Across the junction between two different metals there is a potential difference, the size of the potential difference depending on the metals used and the temperature of the junction. A *thermocouple* is just wires of two different metals forming a complete circuit, with consequently two such junctions (Figure 8.17). When both junctions are at the same temperature then the potential differences across each junction are the same and there is no net e.m.f. A difference in temperature between the two junctions, however, produces a net e.m.f. The value of this e.m.f. depends on the two metals concerned and the temperatures of both the junctions. Usually one junction is held at 0 °C and then

$$E = a_1\theta + a_2\theta^2 + a_3\theta^3 + \text{etc.}$$

where a_1, a_2, a_3, etc. are constants with $a_1 > a_2 > a_3$ and θ the temperature in °C. Figure 8.18 shows, for a number of pairs of metals, how the thermoelectric e.m.f. varies with temperature when one junction is held at 0 °C. Tables, giving the thermoelectric e.m.f. at different temperatures, are available for the metals usually used for thermocouples. Table 8.4 lists commonly used thermocouples and their reference letters and Table 8.5 the thermoelectric e.m.f.s.

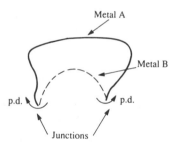

Figure 8.17 A thermocouple

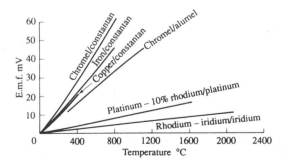

Figure 8.18 Thermoelectric e.m.f.–temperature graphs

Table 8.4 Thermocouples

Type	Materials	Range °C	Sensitivity $\mu V/°C$ in range 0–100 °C
E	chromel – constantan	−200 to 980	63
J	iron – constantan	−200 to 850	53
K	chromel – alumel	−200 to 1300	41
N	nirosil–nisil	−200 to 1300	28
R	platinum – platinum/rhodium 13%	0 to 1400	6
S	platinum – platinum/rhodium 10%	0 to 1400	6
T	copper – constantan	−200 to 370	43

Note: The range is that for constant use.

The tables assume that the temperature of one of the junctions is 0 °C, if other temperatures are used the *law of intermediate temperatures* has to be used to determine the e.m.f. This states that the e.m.f. of a thermocouple with junctions at θ_1 and θ_3 is the algebraic sum of the e.m.f.s of the two thermocouples of the same materials with junctions at θ_1 and θ_2, and θ_2 and θ_3. Thus the e.m.f. at temperature θ when the cold junction is at 0 °C is the sum of the e.m.f. of a thermocouple at temperature θ when the cold junction is at some other temperature and the e.m.f. of the thermocouple at that temperature when the cold junction is 0 °C.

Compensation circuits can be used when the reference junction is not maintained at a constant temperature but allowed to vary with the ambient temperature. The circuit has to supply a potential difference which is the same as the e.m.f. that would be generated by the thermocouple with one junction at 0 °C and the other at the ambient temperature. This can be produced by using a resistance thermometer in a Wheatstone bridge. The bridge is balanced at 0 °C and the out-of-balance potential difference provides the correction potential difference at other temperatures.

For a resistance thermometer element

$$R_\theta = R_0(1 + \alpha\theta)$$

where R_θ is the resistance at temperature θ, R_0 the resistance at 0 °C and α the temperature coefficient of resistance. Hence

change in resistance $= R_\theta - R_0 = R_0\alpha\theta$

The out-of-balance potential difference δV_o for a Wheatstone bridge with the resistance element in one arm and a variable resistor R_2 in a balancing arm is given by (see Chapter 9)

$$\delta V_o = \frac{V_s R_0 \alpha\theta}{R_0 + R_2}$$

V_s is the bridge supply voltage. If the thermocouple e.m.f. E variation with temperature θ can be represented by $E = a\theta$, where a is a constant (the e.m.f. produced per degree change in temperature), then for compensation

Table 8.5 Thermoelectric e.m.f., with reference junction at 0 °C

Temp. °C	Thermocouples, e.m.f. in mV						
	E	J	K	N	R	S	T
-260	-9.795		-6.441	-4.336			-5.891
-240	-9.604		-6.344	-4.277			-5.603
-220	-9.274		-6.158	-4.162			-5.261
-200	-8.824	-7.890	-5.891	-3.990			-4.865
-180	-8.273	-7.402	-5.550	-3.766			-4.419
-160	-7.631	-6.821	-5.141	-3.491			-3.923
-140	-6.907	-6.159	-4.669	-3.170			
-120	-6.107	-5.426	-4.138	-2.807			
-100	-5.237	-4.632	-3.553	-2.407			-3.378
-80	-4.301	-3.785	-2.920	-1.972			-2.788
-60	-3.306	-2.892	-2.243	-1.509			-2.152
-40	-2.254	-1.960	-1.527	-1.023		-0.194	-1.475
-20	-1.151	-0.995	-0.777	-0.518		-0.103	-0.757
0	0.000	0.000	0.000	0.000	0.000	0.000	0.000
20	1.192	1.019	0.798	0.525	0.112	0.113	0.789
40	2.419	2.058	1.611	1.064	0.246	0.235	1.611
60	3.683	3.115	2.436	1.619	0.363	0.365	2.467
80	4.983	4.186	3.266	2.188	0.500	0.502	3.357

100	6.317	5.268	4.095	2.774	0.645	0.645	4.277
120	7.683	6.359	4.919	3.374	0.798	0.795	5.227
140	9.078	7.457	5.733	3.988	0.957	0.950	6.204
160	10.501	8.560	6.539	4.617	1.121	1.109	7.207
180	11.949	9.667	7.338	5.258	1.290	1.273	8.235
200	13.419	10.777	8.137	5.912	1.465	1.440	9.286
220	14.909	11.887	8.938	6.577	1.643	1.611	10.360
240	16.417	12.998	9.745	7.254	1.826	1.785	11.456
260	17.942	14.108	10.560	7.940	2.012	1.962	12.572
280	19.481	15.217	11.381	8.636	2.202	2.141	13.707
300	21.033	16.325	12.207	9.340	2.395	2.323	14.860
320	22.597	17.432	13.039	10.053	2.591	2.506	16.030
340	24.171	18.537	13.874	10.772	2.790	2.692	17.217
360	25.754	19.640	14.712	11.499	2.991	2.880	18.420
380	27.345	20.743	15.552	12.233	3.194	3.069	19.638
400	28.943	21.846	16.395	12.972	3.399	3.260	20.869
420	30.546	22.949	17.241	13.717	3.607	3.452	
440	32.155	24.054	18.088	14.467	3.817	3.645	
460	33.767	25.161	18.938	15.222	4.029	3.840	
480	35.382	26.272	19.788	15.981	4.241	4.036	
500	36.999	27.388	20.640	16.744	4.455	4.234	
550	41.045	30.210	22.772	18.668	5.004	4.732	
600	45.085	33.096	24.902	20.609	5.563	5.237	
650	49.109	36.066	27.022	22.564	6.137	5.751	

continued

Table 8.5 (continued)

Temp. °C	E	J	K	Thermocouples, e.m.f. in mV N	R	S	T
700	53.110	39.130	29.128	24.526	6.720	6.274	
750	57.083	42.283	31.214	26.491	7.315	6.805	
800	61.022	45.498	33.277	28.456	7.924	7.345	
850	64.924	48.717	35.314	30.417	8.544	7.892	
900	68.783	51.875	37.325	32.370	9.175	8.448	
950	72.593	54.949	39.310	34.315	9.816	9.012	
1000	76.357	57.942	41.269	36.248	10.471	9.585	
1050		60.877	43.202	38.169	11.138	10.165	
1100		63.777	45.108	40.076	11.817	10.754	
1150		66.664	48.985	41.966	12.503	11.348	
1200		69.536	48.828	43.836	13.193	11.947	
1250			50.633	45.682	13.889	12.550	
1300			52.398	47.502	14.582	13.155	
1350			54.125		15.276	13.761	
1400					15.969	14.368	
1450					16.663	14.973	
1500					17.536	15.576	
1550					18.043	16.176	
1600					18.727	16.771	
1650					19.409	17.360	
1700					20.090	17.942	
1750						18.504	

$$a\theta = \frac{V_s R_0 \alpha \theta}{R_0 + R_2}$$

$$\frac{R_2}{R_0} = \frac{V_s \alpha - a}{a}$$

The *law of intermediate metals* can be stated as: a thermocouple circuit can have other metals in the circuit and they will have no effect on the thermoelectric e.m.f. provided all their junctions are at the same temperature.

Piezo-electric transducers

15 Piezo-electric force transducer

When certain crystals are subject to tension or compression forces charges appear on their surfaces. This effect is called *piezo-electricity*. Examples of such crystals are quartz, tourmaline and piezo-electric ceramics, such as lead zirconate-titanate.

The charge produced on the surface of a piezo-electric crystal is directly proportional to the applied force.

$$\text{Charge sensitivity } d = \frac{\text{charge } q}{\text{force } F}$$

To measure the charge electrodes are deposited on opposite faces of a crystal. Such an arrangement forms a capacitor and thus a piezo-electric transducer can be considered to be a charge generator in parallel with a capacitor. Table 8.6 gives data for commonly used piezo-electric crystals.

When a load, e.g. a recorder, is connected across the transducer then the potential difference across the load depends not only on the charge sensitivity of the transducer but its capacitance and also the capacitance of the load and its connecting leads. For this reason the transducer is frequently connected via a charge amplifier to the load. With this arrangement the potential difference across the load depends only on charge sensitivity of the transducer and the capacitance of the charge amplifier. The static transfer function of the system is then

$$\text{static transfer function} = \frac{\text{change in p.d. across load } \Delta V_L}{\text{change in force } \Delta F}$$

$$= \frac{\text{charge sensitivity } d}{\text{amplifier capacitance } C_A}$$

Table 8.6 Piezo-electric crystals

Material	Charge sensitivity pC/N	Relative permittivity
Leadmetaniobate	80	250
Lead zirconate-titanate	250	1500
Quartz	2.3	4.5
Tourmaline	1.9 or 2.4*	6.6

* The value depends on the direction of the force relative to the crystal axes.

The dynamic transfer function is that of a second order system.

$$\frac{\Delta V_L}{\Delta F}(s) = \frac{(d/C_A)}{(1/\omega_n{}^2)s^2 + (2\zeta/\omega_n)s + 1}$$

where ω_n is the natural frequency and ζ the damping ratio. Typically ω_n is about $100 \, ks^{-1}$ and ζ about 0.01.

Piezo-electric transducers are used for the measurement of force, pressure, acceleration and vibration. When used for acceleration or vibration the arrangement consists of a mass m held in contact with the surface of a piezo-electric crystal by a spring. When there is an acceleration the mass exerts a force F on the crystal, where $F = ma$. Hence for static conditions, since for the crystal

$$\text{charge sensitivity} = \frac{\text{change in charge}}{\text{change in force}}$$

then

$$\text{charge sensitivity} = \frac{\text{change in charge}}{\text{mass} \times \text{change in acceleration}}$$

For dynamic conditions the system is a second order system as described earlier.

16 Piezo-electric ultrasonic transmitter

The inverse piezo-electric effect is also possible, i.e. a potential difference applied across a crystal in producing charges on opposite crystal faces causes it to contract or expand. By using a high frequency alternating potential difference high frequency oscillations of the crystal are produced. These will result in pressure waves being produced in a medium adjacent to the crystal faces. This is the basis of the ultrasonic generator.

Photovoltaic transducers

17 Photodiodes

A *photodiode* is just a conventional diode in a housing that lets light reach the semiconductor junction. In use the diode is generally reverse biased, the current through the diode then being proportional to the light intensity at the junction. The output is normally taken as the potential difference across a resistor in series with the diode (Figure 8.19). Photodiodes have a lower light sensitivity than photoconductive cells (see earlier this chapter) but a faster response time. Silicon diodes are most efficient at wavelengths of about $0.8 \, \mu m$, with GaAs, GaInAsP and CdTe about 1.1 to $1.5 \, \mu m$ and metal–semiconductor diodes such as Ag–ZnS and Au–Si 0.3 to $0.7 \, \mu m$.

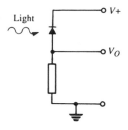

Figure 8.19 Photodiode connections

The *p–i–n photodiode* has a wide layer of intrinsic material between the p and n-type materials in the junction and such a diode has a very fast response time, of the order of nanoseconds. The *avalanche photodiode* is operated at a high reverse bias in the breakdown mode and also gives a fast response time.

18 Photoemissive devices

When light is incident on certain materials electrons are emitted. A *vacuum photoemissive cell* just consists of a light sensitive material in the form of a half cylinder as the cathode and a rod anode in a vacuum tube (Figure 8.20(a)). With the anode maintained positive with respect to the cathode a current flows which is related to the intensity of the light falling on the cathode. The materials used for the photocathode determine the sensitivity of the tube and the wavelengths to which it responds. A commonly used material, referred to as a bialkali type is formed from potassium and caesium or sodium and potassium. Such materials are sensitive to wavelengths in the region of about 200 nm to 700 nm.

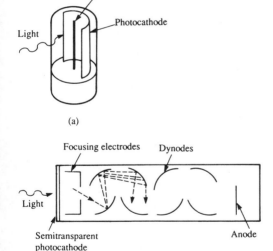

Figure 8.20 (a) Photoemissive cell (b) photomultiplier

The *photomultiplier* (Figure 8.20(b)) has a semitransparent photocathode. The electrons emitted from this cathode are focused onto the first dynode, it being maintained at a positive potential with respect to the cathode and coated with a material such as BeO, CsSb or GaP. The electrons hitting the dynode cause secondary emission with the result that more electrons are emitted. With GaP coatings for each electron hitting the dynode there can be 50 emitted. These then travel to the second dynode where yet further multiplication of the electrons occurs. Eventually after a number of such encounters with dynodes the electrons reach the anode. As a consequence of this large internal amplification, photomultiplier tubes can be used to detect very low levels of illumination.

Electrochemical transducers

19 Ion selective electrodes

When an ion selective electrode is placed in a solution it will give a response which depends mainly on the concentration of a single type of ion in the solution. Figure 8.21(a) shows an electrode which responds to hydrogen ions. The ion selective electrode is used with a suitable standard reference electrode and the potential difference between the two measured with either a high input impedance voltmeter or a buffer amplifier with high input impedance. A commonly used reference electrode is the calomel electrode (Figure 8.21(b)).

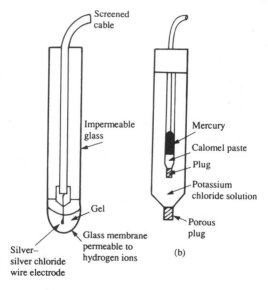

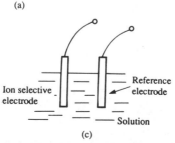

Figure 8.21 (a) Hydrogen ion electrode (b) calomel reference electrode (c) the basic measurement arrangement

The measurement of the concentration of hydrogen ions in a solution is used to give a measure of the acidity or alkalinity of the solution using the pH scale.

$$pH = \log_{10}\left(\frac{1}{\text{hydrogen ion concentration}}\right)$$

A pH value of 7 is neutral, >7 alkaline, <7 acidic.

Elastic transducers

Elastic transducers can all be considered to behave like the mass–spring–damper system described in Chapter 4 and be represented by a second order transfer function.

20   Proving ring

A proving ring is a steel ring which deforms from its circular shape under the action of forces. The amount of deformation is a measure of the forces and can be measured by a dial test indicator gauge, as in Figure 8.22, or by strain gauges attached to the ring. Proving rings are capable of high accuracy and are used for forces in the range 2 kN to 2000 kN.

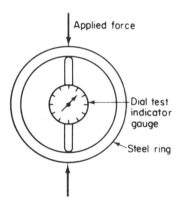

Figure 8.22   Proving ring

21   Load cells

The deformation of a cylinder or a box under the action of forces can be used as a measure of the forces. Such a system is known as a *load cell* (Figure 8.23). Typically they are used for forces in the range 500 N to 6000 kN.

The deformation of the load cell is frequently measured by means of strain gauges (see earlier this chapter). Four identical gauges are often used, each of them forming one arm of a Wheatstone bridge. If the load cell is subject to a compressive load then gauges R_1 and R_3 are in compression, the strain being $-(F/AE)$, where F is the force applied to the cell, A its cross-sectional area and E the tensile modulus of the cell material. Strain gauges R_2 and R_4 are in tension, the strain being $+(vF/AE)$, where v is Poisson's ratio for the cell material. Initially all

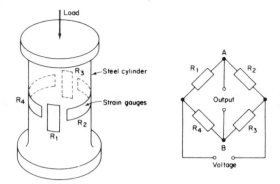

Figure 8.23 A load cell

the gauges have the same resistance R. The effect of the load is to change the resistances.

$$R_1 = R_3 = R - \frac{RGF}{AE}$$

$$R_2 = R_4 = R + \frac{RGvF}{AE}$$

where G is the gauge factor. The out-of-balance potential difference for the bridge is (see Chapter 9)

$$E_{Th} = V_s\left(\frac{R_1}{R_1 + R_4} - \frac{R_2}{R_2 + R_3}\right)$$

$$= V_s\left[\frac{(1 - GF/AE) - (1 + GvF/AE)}{(1 - GF/AE) + (1 + GvF/AE)}\right]$$

Since $(GF/AE) \ll 1$ then

$$E_{Th} = \frac{V_s GF}{2AE}(1 + v)$$

The use of four identical gauges eliminates the effect of temperature since each arm of the bridge is equally affected by any such change.

22 Diaphragms

The movement of the centre of a circular diaphragm when there is a pressure difference between its two sides is the basis of a pressure gauge (Figure 8.24). The amount of movement with a plane diaphragm is

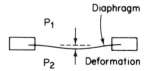

$P_1 - P_2 =$ pressure difference

Figure 8.24 A diaphragm transducer

fairly limited, however greater movement is possible with corrugations in the diaphragm.

23 Capsules and bellows

A capsule (Figure 8.25) can be considered to be just two diaphragms back-to-back, with a bellows just being effectively a stack of capsules.

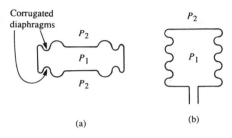

Figure 8.25 (a) A capsule (b) bellows

24 Bourdon tubes

The *Bourdon tube* is an almost rectangular or elliptical shaped tube made from materials such as stainless steel or phosphor bronze. In one form the tube is C-shaped (Figure 8.26). When the pressure inside the tube increases the C opens out, thus the displacement of the closed end of the C becomes a measure of the pressure. A helical form of the tube gives greater deflections. In another version the tube is twisted, the pressure change causing the tube to untwist.

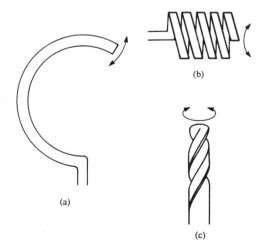

Figure 8.26 Forms of Bourdon tube

Pneumatic transducers

25 Flapper-nozzle

Figure 8.27 shows the basic form of the flapper-nozzle pneumatic transducer. Air at a constant gauge pressure P_s, i.e. the pressure above the atmospheric pressure, flows through the orifice and escapes through the nozzle into the atmosphere. When the flapper closes off the nozzle, i.e. $x = 0$, then no air escapes and the pressure P between the orifice and the nozzle is P_s. When x increases the pressure drops and so becomes a measure of the displacement of the flapper.

The relationship between the pressure P and the displacement x is non-linear and is given by

$$P = \frac{P_s}{1 + 16(d_n^2 x^2 / d_o^4)}$$

where d_n is the diameter of the nozzle and d_o the diameter of the orifice. The transducer has high sensitivity but a small range of measurement, typically ± 0.05 mm.

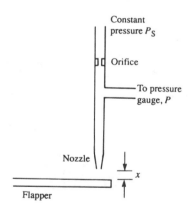

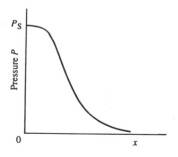

Figure 8.27 Flapper–nozzle

Differential pressure transducers

26 Flowmeters

When a fluid flows from a wider diameter to a narrower diameter pipe its velocity increases and the pressure drops. If the fluid is incompressible, i.e. the density does not change when the pressure changes, Bernoulli's equation gives for a horizontal pipe

$$\frac{v_1^2}{2g} + \frac{P_1}{\rho g} = \frac{v_2^2}{2g} + \frac{P_2}{\rho g}$$

where v_1 is the fluid velocity, P_1 the pressure at the pipe, v_2 the velocity, P_2 the pressure at the constriction, and ρ the fluid density. The mass of fluid flowing through the wider section per second must equal the mass flowing per second through the narrower section. Hence, since the density does not change, the volume of fluid Q passing through the wide section per second must equal the volume passing through the constriction. Hence

$$Q = A_1 v_1 = A_2 v_2$$

where A_1 is the cross-sectional area of the tube and A_2 that at the constriction. Hence

$$Q = \frac{A_2}{\sqrt{[1 - (A_2/A_1)^2]}} \sqrt{[2(P_1 - P_2)/\rho]}$$

In practice, because the flow is not frictionless and the cross-sectional areas of the moving fluid may not be the same as the pipe, the above equation is only an approximation and is modified by a correction factor C.

$$Q = \frac{CA_2}{\sqrt{[1 - (A_2/A_1)^2]}} \sqrt{[2(P_1 - P_2)/\rho]}$$

There are a number of forms of flowmeter based on the measurement of the pressure difference between the flow in the wide cross-section and narrow cross-section of a tube, the different forms being due to the ways by which the narrow cross-section is produced.

The *variable area flowmeter* depends on the same principle as the flowmeters described above, but instead of measuring the pressure difference between the wide and narrow sections of the tube, the size of the narrow section is varied to give a constant pressure difference.

Mechanical transducers

27 Turbine flowmeter

The turbine flowmeter (Figure 8.28) consists of a multi-bladed rotor that is supported centrally in the pipe along which the flow occurs. The angular velocity ω of the rotation of the rotor is proportional to the volume flow rate Q.

$$\omega = kQ$$

where k is a constant which depends on the form of rotor blades. The rate of revolution of the rotor can be determined using a variable reluctance tachogenerator (see item 10 in this chapter). This produces an alternating e.m.f. E.

$$E = N\phi_a n\omega \sin n\omega t$$
$$= N\phi_a nkQ \sin nkQt$$

The alternating e.m.f. has a maximum value of $N\phi_a nkQ$ and frequency $(nkQ/2\pi)$. See item 10 for explanation of the symbols.

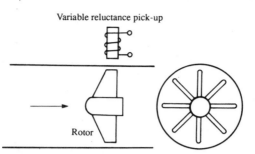

Figure 8.28 Turbine flowmeter

28 Incremental shaft encoder

The angular position of a shaft can be determined using an incremental shaft encoder. It consists of a disc which rotates along with the shaft, the form of the disc depending on the transducer used with it. In one form the disc has a series of windows through which a beam of light can pass (Figure 8.29). The beam of light falls on a light sensitive transducer which gives an electrical output. Rotation of the shaft means that a series of light pulses and hence electrical pulses are produced. The number of pulses produced since some datum position determines the angular position of the shaft. If the rate of rotation of the shaft is required then the number of pulses produced per second is determined.

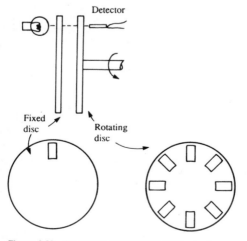

Figure 8.29 An optical incremental shaft encoder

29 Digital shaft encoder

With the incremental shaft encoder the number of pulses has to be counted to give the angular movement and hence angular position. The digital shaft encoder, however, gives an output in the form of a binary number of several digits. Each angular position of the shaft has its own unique binary code. Thus the binary number provides an absolute measurement of the angular position of the shaft. With the optical form (Figure 8.30) the rotating disc has four concentric circles of slots and four sensors to detect the light pulses. The slots are arranged in such a way that the sequential output from the sensors is a number in the binary code.

Other forms of producing the pulses are used. With an electrical method the disc is made of insulating material and has a metallic pattern of segments printed on it. A d.c. voltage is connected to the metallic pattern by means of a brush and the segments of metal can be

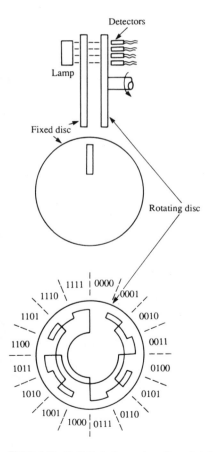

Figure 8.30 Optical shaft encoder with coded disc

Table 8.7 Decimal, binary and Gray code numbers

Decimal	Binary	Gray
0	0000	0000
1	0001	0001
2	0010	0011
3	0011	0010
4	0100	0110
5	0101	0111
6	0110	0101
7	0111	0100
8	1000	1100
9	1001	1101
10	1010	1111
11	1011	1110
12	1100	1010
13	1101	1011
14	1110	1001
15	1111	1000

detected by means of brushes. The electromagnetic form involves a metallic disc with the pattern produced by regions of high and low permeability material. The sensors used are then variable reluctance tachogenerators (see item 10 in this chapter).

Figure 8.30 shows a coded disc which gives an output in the conventional binary code. Problems can however occur if in machining the disc the edges of the various windows are not exactly aligned with each other. Any misalignment will mean that, as the disc rotates, the output from each track will switch at slightly different instants of time and so incorrect binary numbers will occur during that time. This can be overcome by adding an extra outer track, known as the anti-ambiguity track. This consists of small windows which align with the centres of the windows in the other tracks and indicate when the inner tracks are to be read. An alternative method is to use a special form of the binary code called the *Gray code*. With this code only one binary digit changes in moving from one angular position to the next. Table 8.7 shows the relationship between decimal, conventional binary and Gray code numbers.

Vibrating transducers

30 Vibrating wire force transducer
The frequency f with which a taut wire freely vibrates is given by

$$f = \frac{1}{2L} \sqrt{\left(\frac{T}{m}\right)}$$

where L is the vibrating length, T the tension in the wire and m the mass per unit length. This is the basis of a force, or strain, transducer. A wire is clamped at one end and the force to be measured applied to the other. When the force changes then the tension in the wire changes and hence its natural frequency of vibration. This change in frequency can be determined by means of a variable reluctance pick-up.

31 Vibrating cylinder or tube
The natural frequency of vibration of a tube depends on its mass per unit length and any mass which is constrained to vibrate with the tube. Thus the frequency for a tube through which fluid flows will depend on the density of the liquid and can be used for its measurement.

9 Signal converters

Table 9.1 gives a breakdown of the signal converters discussed in this chapter in terms of the forms of their inputs and outputs.

Wheatstone bridge

1 Balanced bridge

The resistances in the arms of the Wheatstone bridge (Figure 9.1) are so adjusted that the output potential difference is zero and a galvanometer connected between the output terminals indicates zero current. The bridge is said to be balanced. When this occurs the potential at B must equal that at D. This means that the potential difference between A and B must equal that between A and D.

$$I_1 R_1 = I_1 R_3$$

Also, the potential difference between B and C must equal that between D and C. Since at balance there is no current through BD then the current through R_2 must be I_1 and that through R_4, I_2. Thus

$$I_1 R_2 = I_2 R_4$$

Hence

$$I_1 R_1 = I_2 R_3 = (I_1 R_2/R_4)R_3$$

$$\frac{R_1}{R_2} = \frac{R_3}{R_4}$$

An unknown resistance R_1 can thus be determined by adjusting, say R_2, to obtain a balance when the ratio R_3/R_4 is maintained constant.

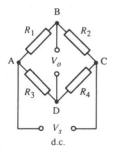

Figure 9.1 The Wheatstone bridge

2 Potential difference output

With no load, i.e. infinite resistance, across the output BD of a Wheatstone bridge (Figure 9.1) the potential drop across the resistor R_1, i.e. V_{AB}, is the fraction $R_1/(R_1 + R_2)$ of the supply voltage V_s.

$$V_{AB} = \frac{V_s R_1}{R_1 + R_2}$$

Similarly, the potential difference across R_3, i.e. V_{AD}, is

$$V_{AD} = \frac{V_s R_3}{R_3 + R_4}$$

Table 9.1 Signal converters

Input	Form of signal Output	Signal converter
Alternating *I* or *V*	Direct *I* or *V*	17
Alternating *V*	Divided alternating *V*	18
Alternating *V*	Modulated *V*	24
Analogue signal	Digital signal	25
Analogue signal	Sampled and held	27
Angular displacement	Amplified angular displacement	11
Angular displacement	Linear displacement	12
Capacitance	Capacitance/Resistance/Inductance	4
Capacitance	Potential difference	5, 6
Charge	Potential difference	13
Current	Scaled current	15, 19
D.c. signal	Modulated signal	23
Digital signal	Analogue signal	26
Displacement	Amplified displacement	10
E.m.f.	Linear or angular displacement	7, 9
Frequencies, band of	Frequencies, all	21, 22
Inductance	Capacitance/Resistance/Inductance	4
Inductance	Potential difference	5, 6
Inputs, many	Sampled	28
Non-linear p.d.	Linear p.d.	14

Potential difference	Linear or angular displacement	7, 9
Potential difference	Amplified potential difference	13
Potential difference	Divided potential difference	18
Potential difference	Scaled potential difference	16, 19
Power	Scaled power	20
Resistance	Potential difference	2, 3
Resistance	Resistance	1, 3
Thermoelectric e.m.f.	Linear or angular displacement	8

The output potential difference V_o is the difference in potential between points B and D.

$$V_o = V_{AB} - V_{AD} = V_s\left(\frac{R_1}{R_1 + R_2} - \frac{R_3}{R_3 + R_4}\right)$$

This equation gives the balance condition equation when V_o is equated to zero.

The relationship between the resistance R_1 and the output voltage V_o is non-linear. However, the greater the value of the ratio R_3/R_4 or R_1/R_2 the more linear the relationship. For a ratio value of 10 or more the relationship is reasonably linear.

A change in resistance from R_1 to $R_1 + \delta R_1$ gives a change in output from V_o to $V_o + \delta V_o$, where

$$(V_o + \delta V_o) - V_o = V_s\left(\frac{R_1 + \delta R_1}{R_1 + \delta R_1 + R_2} - \frac{R_1}{R_1 + R_2}\right)$$

However, if δR_1 is much smaller than R_1 then

$$\delta V_o = \frac{V_s \delta R_1}{R_1 + R_2}$$

Under such conditions the change in output potential difference is proportional to the change in resistance of R_1.

If there is a finite resistance load R_L across the output then a current I_L will flow through the load. This current can be determined with the Thévenin equivalent circuit (Figure 9.2). The Thévenin voltage V_{Th} is the open circuit output voltage V_o derived above. Thus

$$V_{Th} = V_s\left(\frac{R_1}{R_1 + R_2} - \frac{R_3}{R_3 + R_4}\right)$$

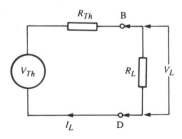

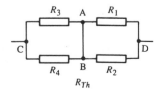

Figure 9.2 Thévenin equivalent circuit

The Thévenin resistance R_{Th} is the resistance seen at points B and D of the bridge and is, assuming the d.c. supply has negligible internal resistance,

$$R_{Th} = \frac{R_1 R_2}{R_1 + R_2} + \frac{R_3 R_4}{R_3 + R_4}$$

The current I_L is thus

$$I_L = \frac{V_{Th}}{R_{Th} + R_L}$$

The potential difference across the load V_L is thus

$$V_L = I_L R_L$$
$$= \frac{V_{Th} R_L}{R_{Th} + R_L}$$

3 Compensated bridge

Compensation for the effects of temperature on the resistance of the leads to a resistance thermometer (see Chapter 21) can be achieved using either a three-lead or four-lead arrangement. Figure 9.3 shows the three-wire arrangement. Lead 1 is in series with the R_3 resistor while lead 3 is in series with the platinum resistance coil R_1. Lead 2 is the connection to the power supply. Any change in lead resistance affects all three leads equally so that changes in lead resistance occur equally in two arms of the bridge and will cancel out if R_1 and R_3 are the same resistance. Figure 9.4 shows the four-wire arrangement. This involves a duplicate set of leads being incorporated in another bridge arm so that any temperature induced resistance changes of the leads cancels out.

The electrical resistance strain gauge (see item 3 Chapter 8) is another transducer where compensation has to be made for temperature effects, the gauge resistance being changed both by changes in strain

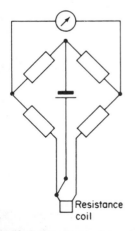

Figure 9.3 Three–wire arrangement for a resistance thermometer

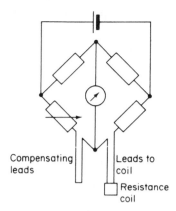

Figure 9.4 Four-wire arrangement for a resistance thermometer

and temperature. Temperature effects can be eliminated by the use of a dummy gauge of identical resistance to the active gauge. The dummy gauge is not subject to the strain but only the temperature. Thus a temperature change will cause both the active and dummy gauges to change resistance by the same amount. The active gauge is mounted in one arm of a Wheatstone bridge and the dummy gauge in an adjacent arm so that the effects of temperature induced resistance changes cancel out (Figure 9.5).

Rather than use a dummy gauge an alternative when both compression and tension are involved, e.g. in a load cell (see item 21 Chapter 8), is to have the compression gauge in one arm of the bridge and the tension gauge in an adjacent arm (see Figure 8.23) so that the temperature induced resistance changes cancel each other out and the strain induced resistance decrease of one gauge and increase for the other produce a greater out-of-balance potential difference than would have occurred with the use of just one of the gauges.

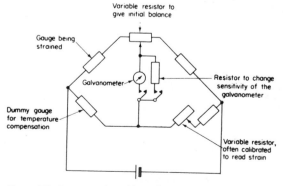

Figure 9.5 Compensation with strain gauges

AC bridges

4 Balanced bridge

Figure 9.6 shows the basic a.c. bridge. The conditions for zero potential difference between A and D, i.e. balance, are that the potential difference across Z_1 must be the same as that across Z_3, in both magnitude and phase, and similarly the potential differences across Z_2 and Z_4 must be equal in both magnitude and phase.

$$I_1 Z_1 = I_2 Z_3$$
$$I_1 Z_2 = I_2 Z_4$$

Hence

$$\frac{Z_1}{Z_2} = \frac{Z_3}{Z_4}$$

This is the basic condition that is applied to any a.c. bridge and has to hold for both the real and imaginary components if the balance is to be valid for both magnitude and phase.

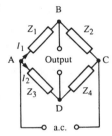

Figure 9.6 Basic a.c. bridge

For a *Maxwell-Wien bridge* (Figure 9.7):

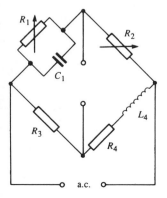

Figure 9.7 Maxwell–Wien bridge

$$\frac{1}{Z_1} = \frac{1}{R_1} + j\omega C_1$$

$$Z_1 = \frac{R_1}{1 + j\omega C_1 R_1}$$

$$Z_2 = R_2$$
$$Z_3 = R_3$$
$$Z_4 = R_4 + j\omega L_4$$

Hence using the balance equation,

$$\frac{Z_1}{Z_2} = \frac{Z_3}{Z_4}$$

$$Z_4 = \frac{Z_2 Z_3}{Z_1}$$

$$R_4 + j\omega L_4 = \frac{R_2 R_3 (1 + j\omega C_1 R_1)}{R_1}$$

For the real parts we have a balance condition of

$$R_4 = \frac{R_2 R_3}{R_1}$$

For the imaginary parts

$$L_4 = R_2 R_3 C_1$$

Thus the resistance R_4 and the inductance L_4 of an inductor can be determined. The procedure is usually to adjust R_2 to obtain the best balance, then R_1 to improve it and then R_2 again and so on until a final balance is obtained. The bridge is used for low Q-value coils.

Figure 9.8 shows three other forms of inductance bridges. For the *Owen bridge* the balance conditions are:

$$L_4 = R_2 R_3 C_1$$

$$R_4 = \frac{R_2 C_1}{C_3}$$

For the *Maxwell bridge* the balance conditions are:

$$L_1 = \frac{R_2 L_3}{R_4}$$

$$R_1 = \frac{R_2 R_3}{R_4}$$

For the *Hay bridge* the balance conditions are:

$$L_1 = \frac{R_2 R_3 C_4}{1 + \omega^2 C_4{}^2 R_4{}^2}$$

$$R_1 = \frac{R_2 R_3 \omega^2 C_4{}^2 R_4{}^2}{1 + \omega^2 C_4{}^2 R_4{}^2}$$

$$Q_1 = \frac{1}{\omega C_4 R_4}$$

Figure 9.9 shows three forms of capacitance bridges. For the *De Souty bridge*:

$$R_1 C_3 = R_2 C_4$$

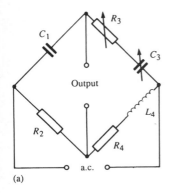

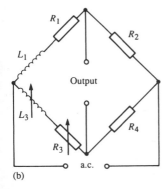

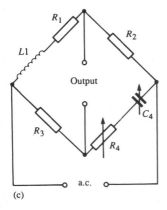

Figure 9.8 (a) Owen bridge (b) Maxwell bridge (c) Hay bridge

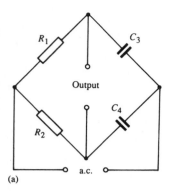

(a)

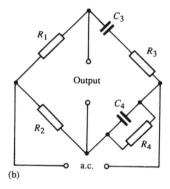

(b)

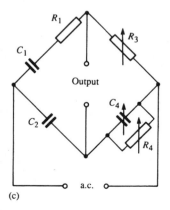

(c)

Figure 9.9 (a) De Souty bridge (b) Wien bridge (c) Schering bridge

For the *Wien bridge*:

$$C_4 = \frac{(R_1/R_2)C_3}{1 + \omega^2 R_3{}^2 C_3{}^2}$$

$$R_4 = \frac{R_2(1 + \omega^2 R_3{}^2 C_3{}^2)}{\omega^2 R_3 R_1 C_3{}^2}$$

For the *Schering bridge*:

$$R_1 = \frac{C_4 R_3}{C_2}$$

$$C_1 = \frac{C_2 R_4}{R_3}$$

loss angle $= \tan \delta = \omega C_4 R_4$

5 Potential difference output

A common form of a.c. bridge involves two resistors (Figure 9.6 with Z_3 and Z_4 resistors) and two reactive impedances, e.g. a pair of inductors or a pair of capacitors. With no load, i.e. infinite load impedance, the output potential difference V_o between B and D, is

$$V_{AD} = E_s\left(\frac{R_4}{R_3 + R_4}\right)$$

$$V_{AB} = E_s\left(\frac{Z_1}{Z_1 + Z_2}\right)$$

$$V_o = V_{AB} - V_{AD} = E_s\left(\frac{Z_1}{Z_1 + Z_2} - \frac{R_4}{R_3 + R_4}\right)$$

For inductors $Z_1 = j\omega L_1$ and $Z_2 = j\omega L_2$, hence

$$V_o = E_s\left(\frac{j\omega L_1}{j\omega L_1 + j\omega L_2} - \frac{R_4}{R_3 + R_4}\right)$$

$$V_o = E_s\left(\frac{L_1}{L_1 + L_2} - \frac{R_4}{R_3 + R_4}\right)$$

For capacitors $Z_1 = 1/\omega C_1$ and $Z_2 = 1/j\omega C_2$, hence

$$V_o = E_s\left[\frac{(1/j\omega C_1)}{(1/j\omega C_1) + (1/j\omega C_2)} - \frac{R_4}{R_3 + R_4}\right]$$

$$V_o = E_s\left[\frac{1}{1 + (C_1/C_2)} - \frac{R_2}{R_3 + R_4}\right]$$

6 Push-pull transducers

A particular use of the bridge outlined above is for *push-pull* forms of transducers. These are transducers involving effectively a pair of capacitors or inductors and the displacement of a plate results in one of the pair increasing and the other decreasing. Figure 9.10 shows a push-pull capacitive transducer. Movement of the central plate by a distance x results in one capacitor increasing its capacitance to $\varepsilon_r\varepsilon_o A/(d-x)$ and the other decreasing it to $\varepsilon_r\varepsilon_o A/(d+x)$. If these two capacitors are used in the bridge shown in Figure 9.9 and the two resistors R_3 and R_4 are equal, then

$$V_o = E_s\left[\frac{1}{1 + (d+x)/(d-x)} - \frac{1}{2}\right]$$

$$= \frac{E_s x}{2d}$$

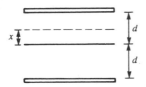

Figure 9.10 Push–pull capacitive transducer

The output voltage from the bridge is thus proportional to the displacement x.

Figure 9.11 shows a push-pull reluctance transducer. Movement of the central plate by a distance x results in the inductance of one inductor increasing and the other decreasing. As indicated in Chapter 8, item 9, one of the inductances will be $L_o/[1 + k(d - x)]$ and the other $L_o/[1 + k(d + x)]$. Hence if these two are the inductors in the bridge described by Figure 9.9, and the two resistors R_3 and R_4 are equal, then

$$V_o = E_s \left[\frac{1/\{1 + k(d-x)\}}{[1/\{1 + k(d-x)\} + 1/\{1 + k(d+x)\}]} - \frac{1}{2} \right]$$

$$V_o = \frac{E_s k x}{2(1 + kd)}$$

The output is thus proportional to the displacement x.

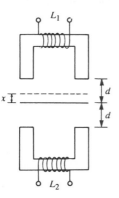

Figure 9.11 Push–pull reluctance transducer

Potentiometer bridges

7 Basic potentiometer circuit

The basic potentiometer circuit (Figure 9.12) has a potentiometer to produce a variable potential difference which is then used to cancel out, i.e. balance, the potential difference being measured. A potential difference is produced across the full length of the potentiometer track and a variable portion of this can be tapped off by movement of a

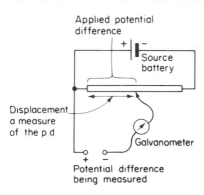

Figure 9.12 Potentiometric measuring system

sliding contact along the track. This potential difference is adjusted until it cancels the potential difference or e.m.f. E being measured, no current then being detected by the galvanometer. If the sliding contact is at a distance L or angle θ from the zero potential difference end of the track then if the track has a uniform resistance per unit length or unit angle

$$E = kL \text{ or } k\theta$$

where k is a constant, in fact the potential difference per unit length or angle of track. This can be determined by repeating the balancing operation with a standard cell of e.m.f. E_s. Then

$$E_s = kL_s \text{ or } k\theta_s$$

where L_s is the balance length or θ_s the balance angle with the standard cell. Hence

$$E = \frac{E_s L}{L_s} \text{ or } \frac{E_s \theta}{\theta_s}$$

With a commercial form of instrument the potentiometer slider moves over a scale marked in volts. The standardization is achieved by setting the pointer to the required value of the standard cell e.m.f. and then adjusting the voltage applied to the potentiometer by adjusting a resistor in series with the voltage source until balance occurs.

The galvanometer is only used to indicate when there is zero current. Because at balance no current is taken from the source being measured, no power is taken. Hence it is a useful method for use with transducers which do not produce much power. The system is essentially an infinite impedance voltmeter.

8 Thermocouple potentiometer bridge

Figure 9.13 shows the form of a potentiometer bridge that can be used with a thermocouple. The e.m.f. from a thermocouple is only millivolts. A resistor is connected in series with the potentiometer track in order to give a small potential difference per unit length or angle of the track. Effectively it is making the track longer. The thermocouple e.m.f. is balanced across a length of track while the standard cell, which has a much larger e.m.f. is balanced across the resistor plus a length of track.

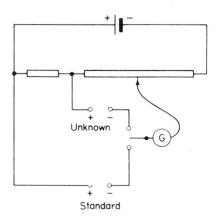

Figure 9.13 Thermocouple potentiometer bridge

9 *Self-balancing potentiometer*

Figure 9.14 shows the basic form of a self-balancing potentiometer bridge. The difference between the potential difference being measured and that across the segment of potentiometer track is fed via an amplifier to a motor. The motor shaft rotates and moves the potentiometer slider across the resistor until the difference between the two signals is reduced to zero. The position of the potentiometer slider is then a measure of the potential difference.

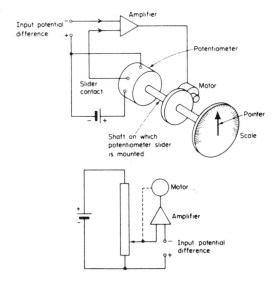

Figure 9.14 A self-balancing potentiometer

Amplifiers

10 *Levers*

Levers are used to change the size of a displacement signal (Figure 9.15). The transfer function of the lever depends on the relative distances from the level pivot point of the application of the input to the lever and of the extraction of the output. Because of similar triangles

$$\frac{\text{input displacement}}{\text{input-pivot distance}} = \frac{\text{output displacement}}{\text{output-pivot distance}}$$

Hence

$$\text{transfer function} = \frac{\text{output}}{\text{input}} = \frac{\text{output-pivot distance}}{\text{input-pivot distance}}$$

Larger magnifications can be produced by a compound lever. With such a lever the output from the first level becomes the input to a second lever. The transfer function of the compound lever is then the product of the transfer functions of the two levers.

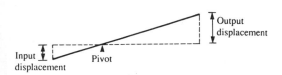

Figure 9.15 Lever

11 *Gear trains*

Gear trains can be used to convert angular displacement of one shaft into a different angular displacement of another. For a two gear wheel system (Figure 9.16) N_I teeth on the input wheel and N_o on the output wheel then one complete revolution of the input shaft means that the output shaft rotates by the fraction (N_I/N_o) of a revolution. Thus the transfer function is

$$\text{transfer function} = \frac{N_I}{N_o}$$

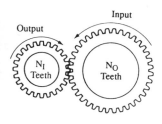

Figure 9.16 A simple gear train

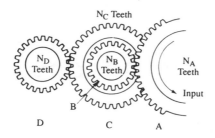

Figure 9.17 A compound gear train

Compound gear trains can be used to give greater magnifications. For the gear train in Figure 9.17, B and C are on the same shaft, and the transfer function is

$$\text{transfer function} = \frac{N_A}{N_B} \times \frac{N_C}{N_D}$$

12 Optical amplification

If a plane mirror is rotated through an angle θ then an incident beam of light has its reflected ray rotated through 2θ (Figure 9.18) and thus an angular displacement has been amplified. Figure 9.19 shows a slightly different arrangement. A source of light placed at the focal point of a convex lens results in parallel rays of light emerging from the lens, the beam of light being said to be *collimated*. If such a beam of light falls on a plane mirror at right angles to its surface then the reflected rays return along the same path as the incident rays and form an image at the same position as the source of light. If however the mirror rotates through an angle θ then the image is formed at a different position to the initial light source. This displacement y is given by

$$\frac{y}{f} = \tan 2\theta$$

where f is the focal length. For the small values of θ used this relationship approximates to

$$\frac{y}{f} = 2\theta$$

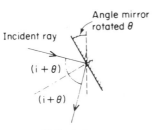

Figure 9.18 Rotation of a plane mirror

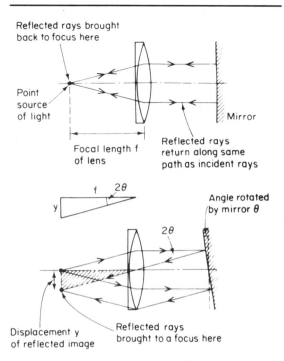

Figure 9.19 Rotation of a plane mirror with collimating lens

This conversion of an angular displacement into a linear displacement of an image is the basis of a number of instruments used for the measurement of angular displacements (see Chapter 11).

13 Operational amplifier

Figure 9.20 shows the connections made to the operational amplifier when it is used as an inverting amplifier. The operational amplifier has a very large transfer function, 100 000 or more, and the change in its output voltage is generally limited to about ± 10 V. The input voltage V_x to the operational amplifier at X must therefore be between $+$ or $-$ 0.0001 V. This is virtually zero and so point X is at virtually earth potential. The potential difference across R_1 is $(V_I - V_x)$, hence since V_x is virtually zero the input potential V_I can be considered to be across R_1. Thus if I_1 is the current through R_1

$$V_I = I_1 R_1.$$

Because the operational amplifier has a very high impedance virtually no current flows through X into it. Hence the current through I_1 flows on through R_f. The potential difference across R_2 is $(V_x - V_o)$, and so the potential difference across R_2 will be virtually $-V_o$. Hence

$$-V_o = I_1 R_f$$

$$\text{transfer function} = \frac{V_o}{V_I} = -\frac{R_f}{R_1}$$

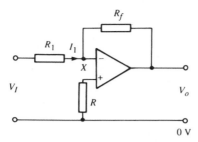

Figure 9.20 Inverting amplifier

The negative sign indicates that the output is 180° out of phase with respect to the input.

Figure 9.21 shows the operational amplifier connected as a non-inverting amplifier. For such an arrangement

$$\text{transfer function} = 1 + \frac{R_f}{R_1}$$

Figure 9.22 shows the operational amplifier connected as a differential amplifier. For such an arrangement

$$V_o = \frac{R_f}{R_1}(V_2 - V_1)$$

Such an amplifier is often used with bridge circuits to amplify the out-of-balance difference in potential.

Figure 9.23 shows the operational amplifier connected as a charge amplifier. For such an arrangement the potential difference across the feedback capacitor is $(V_- - V_o)$ and so the charge on it is $C_f(V_- - V_o)$. But V_- is a virtual earth, hence the charge is $-C_f V_o$. This is the input charge Q, hence

$$V_o = -\frac{Q}{C_f}$$

In an ideal operational amplifier the output would be zero with both the inputs grounded. In practice this is not the case, the operational amplifier behaves as if there was a small potential difference between its inputs, this being called the *differential input offset voltage*. Its value depends on the temperature, thus though it can usually be cancelled out at some particular temperature a change in temperature will result in the changing differential input offset voltage causing the output of the amplifier to drift.

In an ideal operational amplifier the output should depend only on the difference between the signals applied to its two inputs, i.e. $(V_+ - V_-)$ and be independent of the size of the inputs. In practice the output is affected by the common mode voltage V_{CM}, this being the average input voltage. The *common mode rejection ratio* (C.M.R.R.) is defined as

$$\text{C.M.R.R.} = \frac{\text{Differential transfer function } A_d \text{ when } V_{CM} = 0}{\text{Common mode transfer function } A_{CM} \text{ when } V_+ = V_-}$$

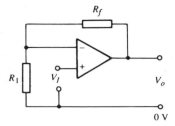

Figure 9.21 Non-inverting amplifier

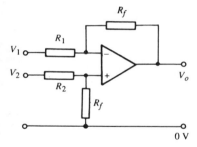

Figure 9.22 Differential amplifier

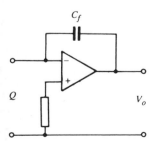

Figure 9.23 Charge amplifier

The output from an operational amplifier is thus

$$V_o = A_d(V_+ - V_-) + A_{CM} V_{CM}$$

$$V_o = A_o \left[(V_+ - V_-) + \frac{V_{CM}}{\text{C.M.R.R.}} \right]$$

With the differential amplifier shown in Figure 9.22 this means that the output becomes

$$V_o = \frac{R_f}{R_1}(V_2 - V_1) + \left(1 + \frac{R_f}{R_1}\right)\frac{V_{CM}}{\text{C.M.R.R.}}$$

The higher the value of the C.M.R.R. the smaller the common mode term and the more closely the expression becomes the ideal one. The C.M.R.R. is usually expressed in decibels

$$\text{C.M.R.R.} = 20\log_{10}(A_d/A_{CM})$$

Typically the C.M.R.R. is about 90 dB.

Further reading: Moore, D. and Donaghy, J., Operational amplifier circuits (Heinemann 1986).

Signal linearization

14 Operational amplifier linearization circuit
An operational amplifier circuit which has been designed to have a non-linear relationship between its input and output can often be used to convert a non-linear input into a linear output. This is achieved by suitable choice of components for the feedback loop, e.g. a diode (Figure 9.24). The diode has a non-linear characteristic and this results in the relationship between the input voltage V_I and the output voltage V_O becoming

$$V_O = C\ln(V_I)$$

with C being a constant. If the input V_I is provided by a transducer which has for its input θ with the relationship between them being

$$V_I = K e^{\alpha\theta}$$

with K and α being constants, then

$$\begin{aligned} V_O &= C\ln(K e^{\alpha\theta}) \\ &= C\ln(K) + C\alpha\theta \end{aligned}$$

Thus the output is directly proportional to the input to the transducer.

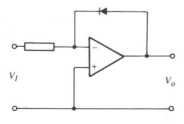

Figure 9.24 Linearization circuit

Current and voltage scaling

15 Shunts

The term shunt is used to describe a resistor connected in parallel with an instrument in order to change the current range of that instrument (Figure 9.25). The potential difference across the instrument is $I_i R_i$ and across the shunt $(I - I_i)R_s$. Since these must be equal,

$$(I - I_i)R_s = I_i R_i$$

$$R_s = \frac{I_i R_i}{I - I_i}$$

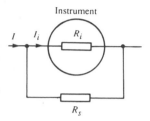

Figure 9.25 Instrument shunt

16 Multipliers

The term multiplier is used to describe a resistor connected in series with an instrument in order to change the voltage range of that instrument (Figure 9.26). The potential difference V is the sum of the potential differences across both the multiplier and the instrument. Hence

$$V = I_i(R_m + R_i)$$

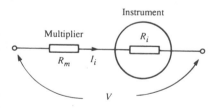

Figure 9.26 Instrument multiplier

17 Rectifiers

Figure 9.27 shows how a basic d.c. meter can be adapted to give a value for a.c. The rectifiers are usually germanium or silicon diodes. In the bridge arrangement shown, a pulsating unidirectional current is produced through the meter. The inertia of the coil of the meter causes it to indicate the average value of the pulses. However the meter is generally calibrated in terms of the root mean square value, on the assumption that the input to the bridge is sinusoidal. The root mean square value is then 1.1 times the average value.

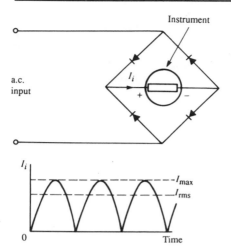

Figure 9.27 Instrument full wave rectifier

18 Voltage dividers

Figure 9.28 shows a number of different forms of voltage dividers. For the resistor chain of Figure 9.28(a) the output potential difference V_o is given by

$$\frac{V_O}{V_I} = \frac{R_1}{R_1 + R_2}$$

Such a divider can be used for d.c. and low frequency a.c. See Chapter 5 for a discussion of the effect of load resistance on the output.

The capacitive divider (Figure 9.28(b)) can be used with a.c. for frequencies from power frequencies to MHz. For the two capacitors in series

$$Q = \left(\frac{C_1 C_2}{C_1 + C_2}\right) V_I$$

Hence since the output is given by $Q = C_2 V_O$, then

$$\frac{V_O}{V_I} = \frac{C_1}{C_1 + C_2}$$

A divider which can span both the low frequency range of the resistive divider and the high frequency range of the capacitive divider is the resistive-capacitive divider shown in Figure 9.28(c). For such an arrangement

$$\frac{V_O}{V_I} = \frac{(R_1/j\omega C_1)/(R_1 + 1/j\omega C_1)}{\left(\dfrac{R_1/j\omega C_1}{R_1 + \{1/j\omega C_1\}}\right) + \left(\dfrac{R_2/j\omega C_2}{R_2 + \{1/j\omega C_2\}}\right)}$$

If the time constant is adjusted to be the same for each resistor–capacitor pair, i.e. $R_1 C_1 = R_2 C_2$, then

$$\frac{V_O}{V_I} = \frac{R_1}{R_1 + R_2}$$

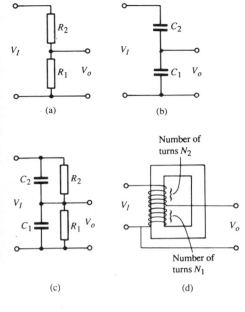

Figure 9.28 Voltage dividers: (a) resistance chain
(b) capacitive (c) resistance–capacitive (d) inductive

An autotransformer is an inductive divider (Figure 9.28(d)) with

$$\frac{V_O}{V_I} = \frac{N_1}{N_1 + N_2}$$

19 Transformers

A *voltage transformer* is a common method of scaling alternating voltages. It consists of concentric windings on a rectangular core, care being taken to minimize core losses. The secondary current is generally small and so

$$\frac{V_s}{V_p} = \frac{N_s}{N_p}$$

where V_s is the secondary voltage, V_p the primary voltage, N_s the number of turns in the secondary and N_p the number of turns in the primary. One of the main advantages of voltage transformers for voltage scaling is that they isolate the primary and secondary circuits. This allows high voltages to be scaled more safely than if the scaling was done by using a multiplier with a meter.

For current scaling with alternating currents, shunts present problems in that the proportion of current that flows through the meter will depend on the impedance of the meter and this varies with frequency. Also, with large currents the shunt resistance needs to be very low and so the shunt and its connections to the meter can become very large and unwieldly. Such problems can be overcome by means of a *current transformer*.

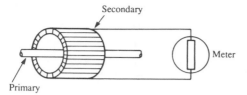

Figure 9.29 Current transformer with bar primary

Current transformers have either just a few turns or a single turn for the primary. The single turn arrangement is referred to as a *bar primary* (Figure 9.29). This primary is in series with the load for which the current is to be measured. The secondary has many turns and has just a meter connected across it. This almost short circuits the secondary and care should be taken never to operate the transformer with the secondary open-circuit since considerable damage can occur. The design of the transformer is such that, to a reasonable accuracy,

$$\frac{I_s}{I_p} = \frac{N_p}{N_s}$$

where I_s is the secondary current and I_p the primary current.

Attenuation

An *attenuator* reduces the power of an input so that the ratio of the input power to the output power is a constant. The *attenuation* A is defined, in decibels, as

$$A = 10 \log_{10}(P_I/P_O)$$

where P_I is the input power and P_O the output power.

For two attenuators in series the output from the first attenuator P_{O1} becomes the input to the second. Thus for the first attenuator

$$A_1 = 10 \log_{10}(P_I/P_{O1})$$

For the second attenuator

$$A_2 = 10 \log_{10}(P_{O1}/P_O)$$
$$A_1 + A_2 = 10 \log_{10}(P_I/P_{O1}) + 10 \log_{10}(P_{O1}/P_O)$$
$$= 10 \log_{10}\left(\frac{P_i}{P_{o1}}\right)\left(\frac{P_{o1}}{P_o}\right)$$
$$= 10 \log_{10}(P_i/P_o)$$

This is however the attenuation A of the system as a whole. Thus

$$A = A_1 + A_2$$

20 Attenuators

In general attenuators are made up of repetitive sections. Figure 9.30 shows a single *T-section* attenuator in a circuit. V_I is the input potential difference, V_O the output. For a symmetrical T-section the input and output impedances are the same, i.e. $R_s = R_L$, and $R_1 = R_2$. For such an arrangement

$$A = 10 \log_{10}\left(\frac{V_I^2/R_s}{V_O^2/R_L}\right) = 20 \log_{10}\left(\frac{V_I^2}{V_O^2}\right)$$

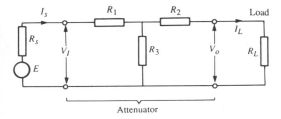

Figure 9.30 T-section attenuator

If we let $N = (V_I^2/V_o^2)$ then $A = 20 \log_{10} N$. For the symmetrical arrangement:

$$R_1 = R_2 = R_L\left(\frac{N-1}{N+1}\right)$$

$$R_3 = R_L\left(\frac{2N}{N^2-1}\right)$$

If the attenuator is required for just matching impedances it is arranged to have the minimum impedance and $R_2 = 0$. The T-section then becomes the *L-section*.

Figure 9.31 shows a *π-section* attenuator. For a symmetrical version of this when the input and output impedances are the same and $R_1 = R_3$, then $A = 20 \log_{10} N$ and

$$R_1 = R_3 = R_L\left(\frac{N+1}{N-1}\right)$$

$$R_3 = R_L\frac{(N^2-1)}{2N}$$

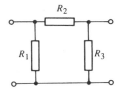

Figure 9.31 π-section attenuator

Filtering

Filtering is the process of removing a certain band of frequencies from a signal and permitting others to be transmitted. A filter is an electrical network which has a frequency dependent transfer characteristic. A filter is said to be *passive* if it is made up of only resistors, capacitors and inductors, and *active* when the filter also involves an operational amplifier. Passive filters have the disadvantage that the current that is drawn by the load can change the frequency characteristic of the filter, this not being the case with an active filter.

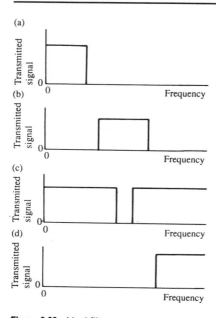

Figure 9.32 Ideal filter responses: (a) a low pass filter (b) a band pass filter (c) a band stop filter (d) a high pass filter

The range of frequencies passed by a filter is known as the *pass band*, the range not passed as the *stop hand* and the boundary between stopping and passing as the *cut-off frequency*. Filters are classified according to the frequency ranges they transmit or reject (Figure 9.32), with a *low pass filter* having a pass band at low frequencies, a *high pass filter* a pass band at high frequencies. A *band-pass* filter allows a particular frequency band to be transmitted, a *band-stop* filter stops a particular band.

21 Passive filters

Two commonly used types of passive filter section are *T-section* and *π-section*. A filter is usually made up of a number of sections, the input impedance of each section being equal to the load impedance for that section. Figure 9.33 shows T-sections, Figure 9.34 π-sections. For the low pass filter section the cut off frequency f_c is

$$f_c = \frac{1}{\pi\sqrt{LC}}$$

For the high pass sections

$$f_c = \frac{1}{4\pi\sqrt{LC}}$$

For the band pass and band stop sections the centre of the band is $\sqrt{f_1 f_2}$, where f_1 is the series arm resonant frequency and f_2 the shunt arm resonant frequency.

Further reading: Niewiadomski, S., Filter handbook (Heinemann-Newnes, 1989).

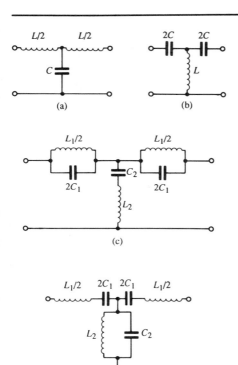

Figure 9.33 T-sections: (a) low pass (b) high pass (c) band stop (d) band pass

22 Active filters

Problems occur with passive filters due to inductors not being resistance-less and with matching successive sections of a filter. These problems can be overcome with filters incorporating operational amplifiers.

Further reading: Hilburn, J. L. and Johnson, D. E. (1973), *Manual of Active Filter Design*, McGraw-Hill.

Modulation

23 Modulation of d.c. signals

Problems that occur with the transmission of low level d.c. signals are drift with operational amplifiers, and external interference. Both these problems can be reduced by modulation, i.e. effectively converting the d.c. signal to an alternating signal. One way is by chopping up the d.c. signal into pulses. The output is then a chain of pulses, the heights of which are related to the d.c. level of the input signal. This process is called *pulse amplitude modulation*. An alternative to this is *pulse width modulation*, often referred to as *pulse duration modulation*. Pulses are

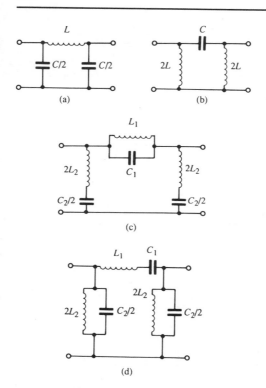

Figure 9.34 π-sections: (a) low pass (b) high pass (c) band stop (d) band pass

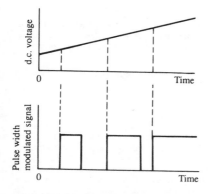

Figure 9.35 Pulse duration modulation

produced at regular intervals, the duration or width of the pulse being proportional to the size of the voltage at each of the times concerned (Figure 9.35).

24 Modulation of a.c. signals

A sinusoidal wave can be represented by

$$v = V \sin(\omega t + \phi)$$

where v is the voltage at some instant of time t, V the maximum value or amplitude of the voltage, ω is the angular frequency ($2\pi \times$ frequency) and ϕ the phase angle. With modulation such a wave is made to act as the carrier for an information signal. *Amplitude modulation* involves the amplitude of a carrier wave being varied according to the size of the information signal (Figure 9.36). *Phase modulation* involves the phase ϕ being varied according to the size of the information signal. *Frequency modulation* involves the ω term being varied according to the size of the information signal (Figure 9.37). Both phase modulation and frequency modulation produce similar effects, a modulated wave with a frequency which relates to the input voltage size.

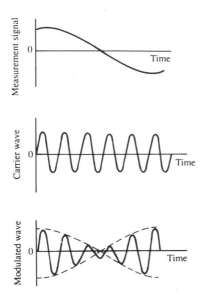

Figure 9.36 Amplitude modulation

Analogue–digital conversion

25 Analogue to digital converters

The input and the output for an analogue to digital conversion (ADC) element are related by

$$V_A \approx V_R(b_1 2^{-1} + b_2 2^{-2} + b_3 2^{-3} + \ldots b_n 2^{-n})$$

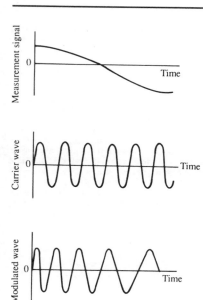

Figure 9.37 Frequency modulation

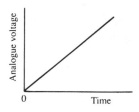

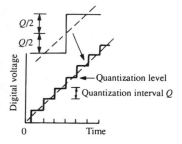

Figure 9.38 Analogue and digital equivalent waveform

where V_A is the analogue input, V_R the reference voltage, $b_1, b_2, b_3 \ldots$ b_n the digital outputs, with n being the number of such outputs which constitute the word representing the analogue signal. Because the output from the converter goes up in steps the equation uses the approximately equals sign $\approx$ (Figure 9.38). The term *quantization* is used for the operation of converting an analogue signal into a number of discrete output steps, each step or voltage level being known as a *quantization level*. The *quantization interval* is the difference in voltage between successive levels. The quantization interval is equal to the output generated by the least significant bit of the binary input word. Because only certain levels are possible there is an error, called the *quantization error* which varies between plus and minus half the quantization interval Q, i.e. $\pm 0.5Q$. This quantization error can be considered to be noise that has been added to the analogue voltage, hence being referred to as *quantization noise*. The quantization error, or quantization noise, is reduced by using a converter controlled by more bits.

The word length determines the *resolution* of the element, i.e. the smallest change in V_A which will result in a change in the digital output. If the analogue to digital converter handles a word of length n bits then a change from 0 to 1 in b_n is the minimum change that can occur and so the resolution is

$$\text{resolution} = V_R 2^{-n}$$

The *maximum value* of the analogue voltage, or *full scale range*, is when all the bits are ones, i.e.

$$\text{maximum } V_A = V_R(1 \times 2^{-1} + 1 \times 2^{-2} + \ldots 1 \times 2^{-n})$$

The value of the bracketed term is $(1 - 2^{-n})$. Hence

$$\text{maximum } V_A = V_R(1 - 2^{-n})$$

For a word length of 4 or more the bracketed term has a value very close to 1 and so the maximum value is virtually the reference voltage.

The *conversion time* of a converter is the time it requires to generate a complete digital word when supplied with the analogue input.

There are a number of forms of analogue to digital converter. The *successive approximations* method involves taking samples of the analogue voltage perhaps 1000 times per second. The converter then supplies a voltage which is compared with the input voltage. The converter voltage goes up in increments, the increments being successively added together until the input voltage is matched. The output from the converter is thus a number representing the increments used. The *ramp* method is essentially a voltage to time converter. At the start of the measurement a ramp voltage starts (Figure 9.39) and is continuously compared with the input voltage. At the start a pulse is generated which opens a gate. The ramp voltage then continues until it becomes equal to the input and another pulse is generated which closes the gate. During the time the gate is open time pulses are counted. The result is that the counted number of pulses is a measure of the input voltage. The above methods give conversions of the analogue signal at some instant of time. Other methods involve integrating the analogue signal over a period of time before converting it to a digital signal. The *dual ramp* method involves a capacitor being charged by the input voltage during a time interval which equals that of 1 cycle of the mains frequency. Then the switch disconnects the input voltage and switches to the reference voltage. The potential difference across the capacitor which resulted from the initial charging

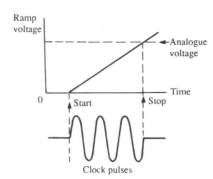

Figure 9.39 Ramp-type analogue to digital converter

is then cancelled at a steady rate and the time taken for it to reach zero is measured. The measurement is by means of counting the number of pulses produced by some 'clock' during that time.

Further reading: Heap, N. W. and Martin, G. S., (1982), *Introductory Digital Electronics*, The Open University Press.

26 Digital to analogue converters

The analogue output V_A of a digital to analogue converter (DAC) is related to the digital input by

$$V_A = V_R(b_1 2^{-1} + b_2 2^{-2} + b_3 2^{-3} + \ldots b_n 2^{-n})$$

where V_R is the reference voltage. This is the maximum voltage the analogue output can have for the word length used. The word length is n bits, with b_1, b_2, b_3 ... b_n being the bits.

The size of the analogue voltage increment when there is a change from 0 to 1 in b_n is

analogue voltage increment $= V_R 2^{-n}$

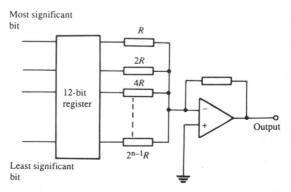

Figure 9.40 Adder digital to analogue converter

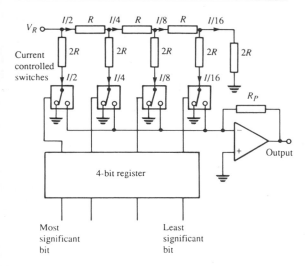

Figure 9.41 Ladder digital to analogue converter

Two basic methods are used for digital to analogue converters, the adder converter and the ladder converter. With the *adder converter* the digital word is loaded into a binary register and its output used to switch a reference voltage on to a series of input resistors of an operational amplifier (Figure 9.40). The values of the resistors are $1R$, $2R$, $4R$, ... $2^{n-1}R$, where n is the word length. The values of the resistors are binary weighted, i.e. the sequence of values is 2^0, 2^1, 2^2, etc. The least significant bit switches the 2^{n-1} resistor on or off, depending whether it is 1 or 0, while the most significant bit switches the R resistor. The adder converter is not generally used for words with more than 6 bits because of the problems of accurately obtaining the high value resistors needed to obtain a suitable range of values. The *ladder converter* uses a resistive ladder (Figure 9.41) and does not have this problem of high resistance resistors since all the resistors have the value R or $2R$. At each junction in the ladder these resistors lead to the current being halved. These currents are switched on or off by the bits. The output of the converter is then determined by the sum of all these currents.

$$\text{Output voltage} = -R_f\left(\frac{I}{2} + \frac{I}{4} + \frac{I}{8} + \ldots\right)$$

Further reading: Heap, N. W. and Martin, G. S., (1982), *Introductory Digital Electronics*, The Open University Press.

27 Sample and hold

With an analogue to digital converter the analogue signal must not change during the time the converter takes to complete the conversion. For this reason a *sample and hold* element is used. It takes a sample of the analogue input and holds it for the analogue to digital converter. Essentially it is a capacitor which, when switched in parallel with the input, is charged up to the analogue voltage. This potential difference is then 'held' until called on by the analogue to digital converter.

The following are terms used in the specifications of sample and hold elements. The *acquisition time* is the time taken for the capacitor to charge up to the value of the input signal. The *aperture time* is the time required for the switch to change state and switch the capacitor in or out. The *holding time* is the length of time the circuit can hold the charge without losing more than a specified percentage of its initial value. The *slew rate* is the maximum rate of output voltage that can be followed.

Further reading: Heap, N. W. and Martin, G. S. (1982), *Introductory Digital Electronics*, The Open University Press.

28 *Analogue multiplexers*
Frequently there is a need for measurements to be made of analogue signals from a number of sources and rather than use separate measurement systems for each a multiplexer can be used with a single sample and hold and analogue to digital converter (Figure 9.42). A *multiplexer* is a switching device which enables each of the inputs to be sampled in turn. A common error with multiplexers is *cross talk*. This is interference occurring between the different inputs.

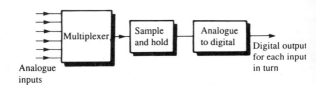

Figure 9.42 Analogue to digital conversion using a multiplexer

10 Display systems

Display systems can be classified into two groups; *indicators* which give an instant visual indication of the process variable, and *recorders* which record the output signal over a period of time and give automatically a permanent record. Both can be subdivided into two groups of devices, *analogue* and *digital*. Table 10.1 shows the classification of the display systems described in this chapter.

Meters

1 The moving coil meter

The meter movement consists of a coil situated in a constant magnetic field which is always at right angles to the sides of the coil no matter what angle the coil has rotated through (Figure 10.1). Consider a coil carrying a current I with vertical sides of length L and horizontal sides length b in a magnetic field which has a uniform flux density B which is always at right angles to the coil. When a current passes through the coil, forces act on the coil sides. The forces on the horizontal sides are in opposite directions and since each side carries the same current these forces cause no motion of the coil, however the forces acting on the vertical sides are also in opposite directions and the same size but they result in a rotation and hence a deflecting torque. The

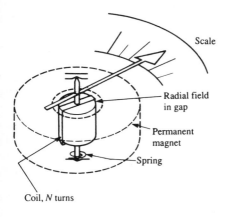

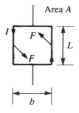

Figure 10.1 The moving coil meter

Table 10.1 Display systems

Display system	Indicator Analogue	Indicator Digital	Recorder Analogue	Recorder Digital
1 Moving coil meter	*			
2 Digital meter		*		
3 Direct reading recorder			*	
4 Direct galvanometric recorder			*	
5 Knife-edge galvan. recorder			*	
6 UV galvanometric recorder			*	
7 Potentiometric recorder			*	
8 X–Y recorder	*			
9 Cathode ray oscilloscope	*			
10 Double beam CRO	*			
11 Sampling CRO				
12 Storage CRO		*	*	
13 Monitors		*		
14 Alphanumeric displays				
15 Dot-matrix printer			*	
16 Direct recording tape recorder			*	
17 Frequency modulated tape recorder				*
18 Digital tape recorder				*

force F acting on a vertical side is BIL. Thus the torque about the central vertical axis of the coil is

$$\text{torque} = BIL \times \tfrac{1}{2}b + BIL \times \tfrac{1}{2}b = BILb$$

Since Lb is the area A of the coil, then the torque is BIA. This is the torque on a single turn of the coil. If there are N turns then the torque is $NBIA$. Since, for a particular galvanometer NB and A will be constant,

$$\text{torque} = K_c I$$

where K_c is a constant for that galvanometer.

This deflecting torque is proportional to the current and causes the coil to rotate. There is, however, an opposing torque generated by springs, the size of this torque depending on the angle through which the coil rotates.

$$\text{Torque due to springs} = K_s \theta$$

where K_s is constant. The coil thus rotates until equilibrium is attained, then

$$K_c I = K_s \theta$$

The angular deflection θ is proportional to the current I.

Moving coil meters generally have resistances of the order of a hundred ohms. Shunts, multipliers and rectifiers (see Chapter 9) are used to convert the basic meter movement to other current and voltage ranges. The accuracy of such a meter depends on such factors as temperature, the presence nearby of magnetic fields or ferrous materials, the way the meter is mounted, bearing friction, inaccuracies in scale marking during manufacture, etc. Parallax errors and errors from interpolating between scale marking can arise in reading the meter. Accuracy is thus of the order of $\pm 0.1\%$ to $\pm 5\%$. The damping of the meter movement is such that the time taken for a steady deflection to be obtained is typically in the region of a few seconds.

2 The digital meter
The digital meter gives its reading in the form of a sequence of digits and hence eliminates parallax and interpolation errors. The meter is an analogue to digital converter connected to a counter. See Chapter 9, item 25, and Chapter 14 for details of the various forms. Accuracies can be as high as $\pm 0.005\%$ with a resistance of the order of 10 MΩ. An important point in considering a digital meter is the rate at which it samples the analogue input signal, the rate depending on the form of ADC used.

Direct reading recorders

3 Direct reading recorder
With the direct reading type of chart recorder a pen is directly moved over a circular chart by the measurement system (Figure 10.2). With a pressure recorder the displacement of the end of a Bourdon tube or bellows may be used, with a temperature recorder the displacement of the end of a bimetallic strip. A circular chart is used and rotates at a constant rate, usually one revolution in 12 hours, 24 hours or 7 days. The pressure or temperature gives displacements out along curved radial lines because the movement of the pen is in the arc or a circle. With respect to time, equal angles are covered in equal intervals of time. This means that the distance moved by the pen in equal intervals of time depends on its distance out from the chart centre. This makes interpolation difficult and there are particular problems in determin-

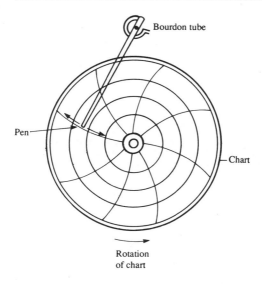

Figure 10.2 Direct-reading recorder

ing values for traces close to the centre of the chart where the radial lines are very close together. After one revolution traces on the chart become superimposed and hence the chart has to be replaced. Simultaneous recording of up to four separate variables is possible. The accuracy is generally of the order of $\pm 0.5\%$ of the full scale deflection of the signal.

Galvanometric recorders

The basis of the galvanometric type of chart recorder is a moving coil galvanometer movement with the resisting torque provided by the twisting of the suspension of the coil. A current through the coil causes it to rotate to an equilibrium position when the torque produced by the current is in equilibrium with the resisting torque. The angular deflection θ of the coil is then proportional to the current (see item 1 and Figure 10.1).

$$K_c I = K_s \theta$$

where K_c is NBA (number of turns × flux density × coil area) and a constant for the galvanometer, K_s is a constant relating the angle of twist of the suspension and its resisting torque.

The above is the static characteristic, i.e. the relationship between the input of the current and the output of a change in angle when a steady current is involved, all transients have ceased and the load across the coil is infinite, i.e. open-circuit. Consider what happens when the current through a galvanometer coil suddenly changes from zero to some value I. The torque due to the current is $K_c I$. The rotation of the coil is opposed by a torque of $K_s \theta$ due to the twisting of the coil suspension. The net torque acting on the coil is thus $(K_c I - K_s \theta)$. This

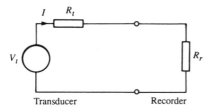

Figure 10.3 Thévenin equivalent circuit

results in an angular acceleration α. Since the net torque is the product of the moment of inertia J and the angular acceleration

$$J\alpha = K_c I - K_s \theta$$

The measurement system can be considered to be represented by the Thévenin equivalent circuit shown in Figure 10.3. Thus

$$I = \frac{V_t}{R_t + R_r}$$

where V_t is the Thévenin equivalent voltage output from the transducer/signal conditioner arrangement, R_t its effective resistance and R_r the effective resistance of the recorder circuit.

The rotation of the coil in a magnetic field results in an induced e.m.f. E. Since for a length L of conductor moving at right angles to a magnetic field $E = BLv$, where v is the linear velocity of the vertical side of the coil, then as $v = r\omega$ where r is the radius of the path and ω the angular velocity $E = BLr\omega$. The radius of the path is however $b/2$ and since there are two sides and n turns

$$E = 2NBL(b/2)\omega = NBA\omega = K_c\omega$$

Thus the voltage in the circuit is $(V_t - K_c\omega)$. Hence the circuit current I is, during the movement of the galvanometer coil,

$$I = \frac{V_t - K_c\omega}{R_t + R_r}$$

Hence substituting for I in the angular acceleration equation

$$J\alpha = \frac{K_c[V_t - K_c\omega)] - K_s\theta}{R_t + R_r}$$

Since $\omega = d\theta/dt$ and $\alpha = d^2\theta/dt^2$,

$$\frac{d^2\theta}{dt^2} + \frac{K_c^2}{J(R_t + R_r)}\frac{d\theta}{dt} + \frac{K_s\theta}{J} = \frac{K_c V_t}{J(R_t + R_r)}$$

This second order differential equation describes the motion of the galvanometer coil when subject to a change in current. The system has a natural frequency of oscillation $\omega_n = 2\pi f_n$ given by

$$\omega_n = \sqrt{(K_s/J)}$$

and a damping factor of

$$\text{damping factor} = \frac{K_c^2}{2(K_s J)^{\frac{1}{2}}(R_t + R_r)}$$

The damping factor involves the effective resistance of the transducer.

Table 10.1   Overshoot and damping factor

Damping factor	% overshoot
1.00	0.0
0.91	0.1
0.82	1.2
0.72	4.0
0.62	8.4
0.50	16.5
0.40	25.0

Thus the damping can be altered by adding resistors in series or parallel with the transducer.

At damping factors which are less than 1.0, the critical damping (see Chapter 4 and Figure 4.3), the deflection overshoots the steady value before settling back to it (see Table 10.1). With damping factors greater than 1.0 the coil just takes a long time to attain the steady value.

When the recorder galvanometer reaches the steady reading both $d^2\theta/dt^2$ and $d\theta/dt$ are zero. Hence

$$\frac{K_s\theta}{J} = \frac{K_c V_t}{J(R_t + R_r)}$$

The steady state voltage sensitivity, is thus

$$\text{voltage sensitivity} = \frac{\theta}{V_t} = \frac{K_c}{K_s(R_t + R_r)}$$

The steady state sensitivity can thus be changed by adding resistors in series or parallel with the transducer. In considering connecting resistors in parallel account has to be taken of not only the change in the effective resistance R_t but also the voltage applied to the recorder. The Thévenin equivalent voltage for the transducer is reduced by a factor of $R_p/(R_p + R_t)$, where R_p is the resistance inserted in parallel with the transducer.

The response of the galvanometer to different frequency sinusoidal inputs depends on the value of the natural frequency and the damping

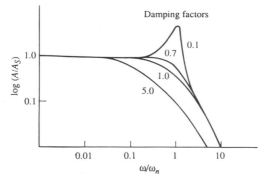

Figure 10.4   Frequency response of a galvanometer

(Figure 10.4). The amplitude A of a signal compared with the value that would pertain under static conditions A_s is given by

$$\frac{A}{A_s} = \frac{1}{[(1 - \omega r^2)^2 + (2\zeta\omega r)^2]^{\frac{1}{2}}}$$

where $\omega_r = \omega/\omega_n$, i.e. the angular frequency relative to the natural frequency. The optimum value of the damping factor to give the greatest bandwidth is about 0.7.

Galvanometers for use at high frequencies thus require a damping factor of about 0.7 and a high natural frequency. The natural frequency can be increased by reducing the moment of inertia J of the coil, i.e. a slim coil with a small breadth. However this reduction also increases the damping factor. Another way by which the natural frequency can be increased is by increasing K_s. This however also affects the sensitivity of the galvanometer. Galvanometers designed for use at high frequencies thus tend to have low sensitivities.

4 Direct galvanometric recorder

There are a number of chart marker mechanisms. With the *direct type* a pen is directly driven by the motion of the galvanometer coil (Figure 10.5). The displacement y of the pen is

$$Y = R \sin \theta$$

where R is the length of the pen arm. The relationship between the

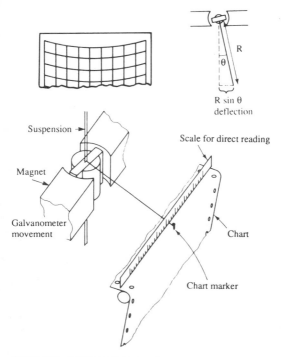

Figure 10.5 Galvanometric recorder

current, which is proportional to θ, and the displacement is thus non-linear. If angular deflections are restricted to less than $\pm 10°$ then the error due to this non-linearity is less than 0.5%. Because the pen moves in an arc rather than a straight line, curvilinear paper has to be used for the plotting and this results in interpolation problems. Accuracies are of the order of $\pm 2\%$ of full scale deflection, input resistances of about $10\,\text{k}\Omega$ and a bandwidth from d.c. to about $50\,Hz$.

5 Knife-edge galvanometric recorder

One way of having rectilinear rather than curvilinear chart is shown in Figure 10.6, the knife-edge recorder. Heat-sensitive paper moves over a knife edge and a heated stylus marks it. The paper may be impregnated with a chemical that shows a marked colour change when heated by contact with the stylus or the stylus burns away temperature sensitive outer layers which coat the paper. The length of trace y produced on the paper by a deflection θ is

$$y = R \tan \theta$$

The relationship between the deflection and the current, which is proportional to θ, is non-linear. If deflections are restricted to less than $\pm 10°$ the non-linearity error is less than 1%. Accuracies are of the order of $\pm 2\%$ of full scale deflection, input resistance of about $10\,\text{k}\Omega$ and a bandwidth from d.c. to about $50\,\text{Hz}$.

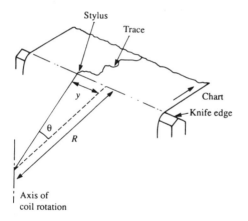

Figure 10.6 Knife edge recorder

6 Ultraviolet galvanometric recorder

The ultraviolet recorder has a small mirror attached to the suspension (Figure 10.7). A beam of ultraviolet light is reflected from the mirror and when the coil rotates the reflected beam is swept across the chart. The chart uses photosensitive paper which on developing shows the trace. A number of galvanometer blocks, 6, 12 or 25, are generally mounted side-by-side in the one magnet so that simultaneous recordings can be made of a number of variables.

The sensitivity depends on the bandwidth required, this as indicated earlier determining the characteristics of the galvanometer mounting. With a bandwidth of d.c. to about $50\,Hz$ the sensitivity is typically about $5\,cm/mV$, the coil having a resistance of about $80\,ohms$. A bandwidth up to $5\,kHz$ would typically have a sensitivity of about

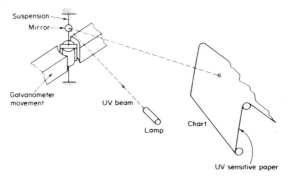

Figure 10.7 Ultraviolet galvanometric recorder

0.0015 cm/mV and a coil resistance of about 40 ohms. The limiting frequency for this type of instrument is about 13 kHz. Accuracy is about $\pm 2\%$ of the full scale deflection.

Closed-loop recorders

7 *Potentiometric recorder*

The potentiometric recorder is a self-balancing potentiometer (Figure 10.8). When there is an input to the motor, the motor shaft rotates and

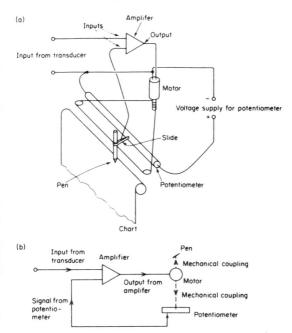

Figure 10.8 Potentiometric recorder

moves the pen. The pen continues in motion as long as there is an input to the motor. However the motion of the pen results in a slider moving along a potentiometer track. This produces a potential difference which is subtracted from the transducer signal by an operational amplifier, the difference signal then being used to operate the motor. The pen thus ends up moving to a position where the result is no difference between the pen and transducer signals.

Potentiometric recorders are more robust than galvanometric recorders; can be multi-channel; have a linear relationship between deflection and transducer input, typically have high input resistances and higher accuracies (about $\pm 0.1\%$ of full scale reading) than galvanometric recorders but considerably slower response times and hence the bandwidth is only d.c. to 1 or 2 Hz. Because of friction there is a minimum current required to get the motor operating, i.e. a dead band. This is usually about $\pm 0.3\%$ of the range.

8 X–Y recorder

The X–Y recorder is used to monitor the relationship between two variables, one of which does not have to be time. The potentiometric recorder can have the pen driven by two motors, one motor controlling the movement in the X-direction and the other in the Y-direction.

Cathode ray tubes

9 Cathode ray oscilloscope

The cathode ray tube (Figure 10.9) consists of an electron gun which produces a focused beam of electrons and a deflection system. In the gun electrons are produced by the heating of the cathode, the number of these electrons which form the electron beam, and determine the brilliance, being determined by a potential applied to the modulator. The electrons are accelerated down the tube by the potential difference between the cathode and the anode with the potential on further electrodes being adjusted to focus and beam so that when it reaches the phosphor coated screen it forms a small luminous spot. The beam is deflected in the Y direction by a potential difference applied between the Y deflection plates and in the X direction by a potential difference between the X deflection plates.

The screen is coated with a phosphor which glows on impact by electrons. The light produced by the phosphor takes a little time to build up and time to decay (Figure 10.10). The time taken for the light output to fall to some specified value of its initial value is known as the *decay time* or *persistence* (Table 10.2).

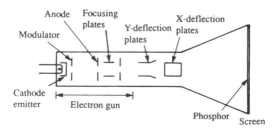

Figure 10.9 Cathode ray tube

Table 10.2 Screen phosphors

Type	Fluorescence	Relative luminance	Decay time to 0.1% in ms	Uses
P.1	yellow–green	50%	95	General purpose. Now mainly replaced by P.31.
P.2	blue–green	55%	120	Suitable for slow varying signals.
P.4	white	50%	20	Used for TV displays.
P.7	blue–white	35%	1500	Long persistence, used for very slow varying signals.
P.11	blue	35%	20	Short persistence, used for photography and fast varying signals.
P.31	green	100%	32	Brightest phosphor, general purpose use.

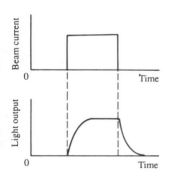

Figure 10.10 Phosphorescence build up and decay time

The function of the *Y deflection plates* is to produce deflections of the electron beam in the vertical direction. A switched attenuator and amplifier enable different deflection factors to be obtained. A general purpose oscilloscope is likely to have deflection factors which vary between 5 mV per scale division to 20 V per scale division. In order that a.c. components can be viewed in the presence of high d.c. voltages a blocking capacitor can be switched into the input line. When the amplifier is in its a.c. mode its bandwidth typically extends from about 2 Hz to 10 MHz, when in its d.c. mode from d.c. to 10 MHz. The input impedance is about 1 mΩ shunted with about 20 pF capacitance.

When an external circuit is connected to the Y-input, problems due to loading and interference can distort the input signal. Interference can be reduced by the use of coaxial connecting cable. However, the capacitance of the coaxial cable and any probe attached to the cable can be enough, particularly at high frequencies, to introduce a relatively low impedance across the input impedance of the oscilloscope and so introduce significant loading effects. A number of probes exist which are designed to increase the input impedance and so avoid this loading problem. A *passive voltage probe* used is a 10-to-1 attenuator, generally a 9 MΩ resistor and variable capacitor in the probe tip (Figure 10.11). This not only reduces the capacitive loading but also the voltage sensitivity (it is an *R–C* voltage divider, see Chapter 9 and Figure 9.28(c)). An *active voltage probe* using an FET can overcome this problem of reduced voltage sensitivity. Another form of probe is the *current probe* (Figure 10.12). This probe can be clamped around the current carrying conductor and does not involve physically inserting any element into the current carrying circuit. The probe is essentially a current transformer.

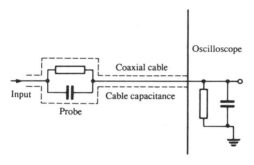

Figure 10.11 Passive voltage probe

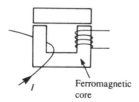

Figure 10.12 Current probe

The purpose of the *X deflection plates* is to deflect the electron beam in the horizontal direction. They are generally used with an internally generated signal which sweeps the beam from left to right at a constant velocity with a very rapid return, i.e. flyback, which is too fast to leave a trace on the screen (Figure 10.13). The constant velocity means that the distance moved in the X direction is proportional to the time elapsed thus giving a horizontal time axis, i.e. a *time base*. A general purpose oscilloscope will have range of time bases from about 1 s per scale division to 0.2 μs per scale division.

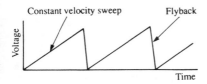

Figure 10.13 Sawtooth waveform for timebase

For a periodic input signal to give rise to a steady trace on the screen it is necessary to synchronize the timebase and the input signal using the *trigger circuit*. The trigger circuit can be adjusted so that it responds to a particular voltage level and also whether the voltage is increasing or decreasing. Hence the trigger circuit responds to particular points on the input waveform (Figure 10.14) and produces pulses which trigger the timebase into action. The timebase sweeps across the screen and so always starts at the same point on the input signal and hence successive scans of the input signal are superimposed.

Further reading: Hickman, I. (1990), *Oscilloscopes* (3rd edn), Butterworth-Heinemann.

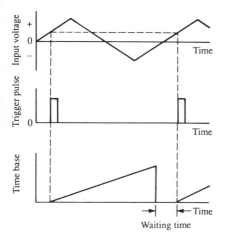

Figure 10.14 Triggering

10 Double beam oscilloscope

Double beam oscilloscopes enable two separate inputs to be observed simultaneously on the screen. This can be by having two independent electron gun assemblies and consequently two electron beams, each beam with its own Y deflection plates but a common set of X deflection plates and so common timebase. Alternatively a single electron gun can be used and the Y deflection plates switched from one input signal to the other each time the timebase is triggered. This is known as the *alternate mode*. Another alternative is to sample the two inputs more frequently, alternating typically from one to the other at about 150 Hz. This is known as the *chopping mode*.

Further reading: Hickman, I., (1990), *Oscilloscopes*, (3rd edn), Butterworth-Heinemann.

11 Sampling oscilloscope

The upper frequency limit of the conventional oscilloscope is of the order of 100 MHz. Higher frequencies can however be used with a sampling oscilloscope. Such an instrument constructs a continuous display from samples taken at different points during a number of cycles of the waveform. The sampling frequency may be one-hundredth of the input frequency and so the upper frequency limit can be extended by a factor of one hundred.

Further reading: Hickman, I., (1990), *Oscilloscopes*, (3rd edn), Butterworth-Heinemann.

12 Storage oscilloscope

With storage oscilloscopes the trace produced remains on the screen after the input signal has ceased, a deliberate action of erasure being necessary to remove it. The *bistable storage tube* (Figure 10.15) has three electron guns. Two of the guns, called the flood electron guns, are on all the time and flood the screen with low velocity electrons. The screen consists of phosphor particles on a dielectric sheet, backed by a conducting layer. The low velocity flood gun electrons cause the phosphor particles to become negatively charged. The phosphor is then in a 'not glowing' state. The writing electron gun emits high velocity electrons, these having sufficient kinetic energy to overcome the repulsion due to the negative charge on phosphor particles and eject electrons from them. These electrons are gathered and conducted away by the conducting layer which backs the phosphor coated sheet. The result of the phosphor particles losing electrons is that they become positively charged. This charge remains, even when the writing gun stops emitting electrons, because the phosphor particle is being bombarded by flood electrons which are accelerated towards it by the positive charge. The resulting kinetic energy is sufficient for the

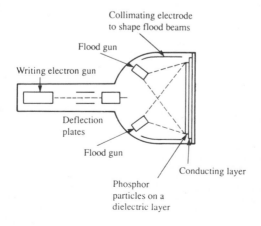

Figure 10.15 Bistable storage cathode ray tube

electron emission to continue. Thus the phosphor is in a 'glowing state' and remains in that condition until deliberate action is taken to 'erase' the charge on the phosphor.

Further reading: Hickman, I., (1990), *Oscilloscopes* (3rd edn), Butterworth-Heinemann.

13 Monitors

A monitor is a display device, involving a cathode ray tube, which can display alphanumeric, graphic and pictorial data. Sawtooth signals are applied to both the X and the Y deflection plates (Figure 10.16). The X input moves the electron beam from left to right across the screen with a rapid flyback. The Y input moves the beam relatively slowly from the top to the bottom of the screen before a rapid flyback to the top of the screen. The result of both these inputs is that the spot zig-zags down the screen. During its travel the electron beam is switched on or off by an input to the modulation electrode and so a 'picture' is painted on the screen. For a stationary picture to remain painted on the screen it is necessary to keep re-energizing the phosphor particles. This type of display is produced by what is called a *refresh-type* of cathode ray tube. An alternative way of retaining the picture on the screen is to use a *storage-type* of cathode ray tube.

A colour monitor has a screen coated with clusters of dots of three different types of phosphor, one to emit red light, one green light and the other blue light. Three electron beams sweep across the screen, one beam for each colour. If the three beams energize all three the appearance is of white light, if just the red phosphor then red light.

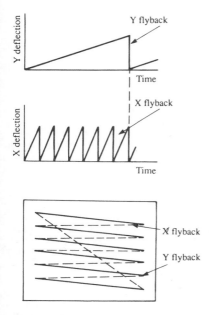

Figure 10.16 Monitor

Alphanumeric displays

The term alphanumeric is a contraction of the two terms alphabetic and numeric and describes the display of the letters of the alphabet, numbers 0 to 9 and decimal points.

14 Alphanumeric display systems

A number of devices can be used as the basis of such displays, the two most common being liquid crystals and light-emitting diodes. A liquid crystal display consists of a film of liquid crystal sandwiched between two transparent electrodes (Figure 10.17). The electrodes are arranged in the form of the segments required for the display. When a potential difference is applied between the electrodes the liquid crystal scatters light which otherwise would have passed straight through it. Thus if light is directed onto the crystal from the front or side the area between the electrodes becomes bright and so the required character becomes visible.

Light emitting diodes emit light when forward biased. By arrangement of the diodes into the arrangement needed to generate the characters, biasing of the appropriate diodes results in the appropriate display.

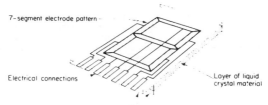

7-segment electrode pattern

Electrical connections

Layer of liquid crystal material

Glass plates with electrode patterns evaporated on

Figure 10.17 Liquid crystal display

A commonly used format for the characters involves seven segments, the various characters being generated by the appropriate selection of segments within the seven (Figure 10.18). Table 10.3 shows how a four-bit binary code input can be used to generate the inputs to switch on the various segments. Another form involves a 7 by 5 or 9 by 7 dot-matrix. The characters being generated by the selection of appropriate dots (Figure 10.19).

Further reading: Marston, R. M., (1988), *Optoelectronics Circuits Manual*, Butterworth-Heinemann.

15 Dot-matrix printer

The dot matrix printer is a commonly used digital printer, giving its output as alphanumeric characters. The print head consists of either nine or twenty-four pins in a vertical line with each pin controlled by an electromagnet which when turned on propels the pin onto an inking ribbon. This impact forces a small blob of ink onto paper behind the ribbon. Characters are formed by moving the print head across the paper and firing the appropriate pins.

Magnetic tape recorders

The magnetic tape recorder consists of a recording head which responds to the input signal and produces corresponding magnetic patterns on magnetic tape, a replay head to convert the magnetic

123

Table 10.3 Seven-segment display

Binary input D C B A	Segments activated a b c d e f g	Number displayed
0 0 0 0	1 1 1 1 1 1 0	0
0 0 0 1	0 1 1 0 0 0 0	1
0 0 1 0	1 1 0 1 1 0 1	2
0 0 1 1	1 1 1 1 0 0 1	3
0 1 0 0	0 1 1 0 0 1 1	4
0 1 0 1	1 0 1 1 0 1 1	5
0 1 1 0	0 0 1 1 1 1 1	6
0 1 1 1	1 1 1 0 0 0 0	7
1 0 0 0	1 1 1 1 1 1 1	8
1 0 0 1	1 1 1 0 0 1 1	9

The segments

Figure 10.18 Seven-segment display

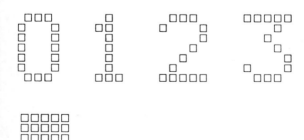

Figure 10.19 7 by 5 dot matrix display

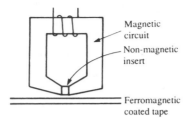

Figure 10.20 Recording/replay head

patterns on the tape to electrical signals, a tape transport system which moves the magnetic tape at a constant speed under the heads, and signal conditioning elements such as amplifiers and filters. The recording head is in the form of a closed loop of ferromagnetic material which has a non-magnetic insert (Figure 10.20). The magnetic tape is a flexible plastic base coated with a ferromagnetic powder. The proximity of the magnetic tape to the nonmagnetic insert means that the magnetic flux detours through the tape. Magnetic flux is produced in the core and the tape by passing an electric current through a coil wrapped round the core. Figure 10.21 shows the relationship between the remanent flux density produced in the tape, i.e. its degree of permanent magnetism, and the magnetizing field strength, this being directly proportional to the current through the coil. Thus a magnetic record is produced on the tape which is related to the current through the recording head coil. The replay head has the same form of construction as the recording head. When a piece of magnetized tape bridges the nonmagnetic gap then magnetic flux is induced in the core. Flux changes in the core induce e.m.f.s in the coil wound round the core and so give an electrical signal output from the coil related to the magnetic record on the tape.

Recorders generally have more than one recording head, the heads being spaced across the width of the tape. Thus several different signals can be simultaneously recorded.

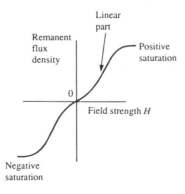

Figure 10.21 Remanent magnetization graph for the tape

16 Direct recording tape recorder

With direct recording the input signal is directly used to produce the remanent magnetism in the tape. The remanent flux density–magnetic field strength graph is non-linear. Thus a current input which fluctuates about the zero gives a distorted variation in remanent flux density in the tape. This distortion can be reduced by adding a steady d.c. current to the signal, i.e. a *d.c. bias*, to shift the signal to more linear parts of the graph (Figure 10.22(a)). An alternative is to add a high frequency a.c. current to the signal, i.e. an *a.c. bias*, and so have an

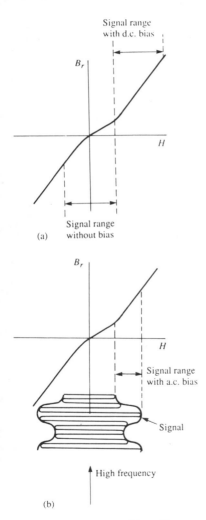

Figure 10.22 (a) d.c. bias (b) a.c. bias

amplitude modulated input (Figure 10.22(b)). The amplitude variations represent the signal and these then occur on the more linear parts of the graph.

For a sinusoidal input to the record head with a frequency f a sinusoidal variation in magnetization is produced along the tape. The time interval for one cycle is $1/f$ and so for a tape moving with a constant velocity v then the distance along the tape taken by one cycle is v/f. This distance is called the *recorded wavelength*.

$$\text{Recorded wavelength} = \frac{v}{f}$$

The minimum size this recorded wavelength can have is the gap width since the average magnetic flux across the gap is then the average of one cycle and so has a zero value. The upper limit to the frequency response of the recorder is thus set by the gap width and the tape velocity. Typically tape velocities range between about 23 mm/s to 1500 mm/s with a gap width of 5 μm, hence upper frequency limit ranges from about 4.6 kHz up to 300 kHz.

For the replay head a sinusoidal variation of flux on the recording tape will produce a sinusoidal variation of flux ϕ in the core.

$$\phi = \phi_m \sin \omega t$$

where $\omega + 2\pi f$, f being the frequency, and ϕ_m is the maximum flux. The output from the replay head is proportional to the rate of change of flux ϕ in the head core, i.e. $d\phi/dt$. Hence

Head output is proportional to $\omega \phi_m \cos \omega t$

Thus the head output depends not only on the flux recorded on the tape but also the frequency. To overcome this the output is amplified by an amplifier with a transfer function which varies with frequency in the way shown in Figure 10.23. The process is called *amplitude equalization*.

A consequence of the replay head output being proportional to the frequency is that for very low frequencies the output may be very small and become comparable with noise picked up by the replay head and so there is a lower frequency limit, about 100 Hz, for which the recorder can be used.

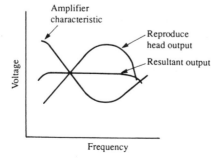

Figure 10.23 Amplitude equalization

17 Frequency modulated tape recorder

With *frequency modulation* the carrier frequency alters in accordance with the fluctuations of the input signal. Because the carrier frequency is high the direct-recording problem of dealing with low frequencies does not occur and so frequently modulation recording can be used down to 0 Hz. The upper frequency limit is, however, less than with direct-recording, typically about a third of the carrier frequency, i.e. in the region 2 to 80 kHz. Frequency modulation tends to give a better signal-to-noise ratio than with direct recording. The control of the tape speed is vital since fluctuations in this can lead to apparent frequency fluctuations.

18 Digital tape recorder

With *digital recording* signals are recorded as a coded combination of bits. A commonly used method is the *nonreturn-to-zero (NRZ)* method. With this system the flux recorded on the tape is either at the positive saturation value or the negative saturation value (see Figure 10.21). No change in flux is used to represent 0 and a change in flux, 1 (Figure 10.24). The output from the replay head depends on the rate of change of flux on the tape and so outputs only occur where the recorded tape has a change of flux, hence the output is a pulse whenever a 1 is recorded. Digital recording has the advantages over analogue recording of higher accuracy and relative insensitivity to tape speed.

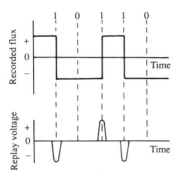

Figure 10.24 Nonreturn-to-zero recording

Part Three
Measurements

11 Chemical composition

This chapter outlines some of the main methods used for chemical analysis. Table 11.1 lists the methods and their main characteristics.

Further reading: Noltingk, B. E. (ed.), (1985), *Jones' Instrument Technology*, vol. 2 (*Measurement of Temperature and Chemical Composition*), (4th edn), Butterworth-Heinemann; Willard H. H., Meritt, L. L., Dean, J. A. and Settle, F.A. (1988), *Instru. 'ental Methods of Analysis*, Wadsworth.

Chromatography

Chromatography is the technique for separating a mixture into its constituent components as a result of a moving phase passing through or over a stationary phase. The sample is introduced into the moving phase and is carried along by it, components in the sample undergoing repeated interactions with the stationary phase until at the end of the process the components have become separated. The mobile phase can be a liquid or gas, the stationary phase a liquid or solid.

1 Paper chromatography

Figure 11.1 shows the apparatus used for paper chromatography. In (a) the liquid descends the paper and in (b) it ascends. The mixture is applied as a spot on the paper and the moving liquid causes the mixture to separate out into a series of spots at different distances from where the sample was applied. The location of these spots can be

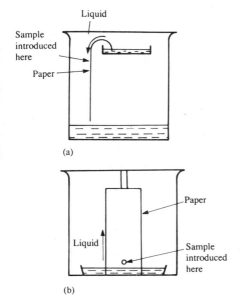

Figure 11.1 Paper chromatography, with (a) descending liquid (b) ascending liquid

Table 11.1 Chemical analysis

Principle	System	Characteristics
Chromatography	1 Paper	Simple laboratory method of separating molecules.
	2 Thin layer	More sensitive than paper.
	3 Liquid	Better reproducibility, resolution and accuracy than thin layer but less sensitive, used for separation of molecules of less volatile and ionic materials.
	4 Gas	Used with volatile materials for separation of molecules.
Electrochemical	5 Conductivity	Measurement of ion concentrations in electrolytes.
	6 Galvanic	Measurement of ionic concentration in electrolytes, including pH.
	7 D.C. polarography	Measurement of ionic concentration.
	8 Anode stripping	More sensitive than polarography, measurement of trace metals.
Spectroscopy	9 UV & visible absorption spec.	Very sensitive and accurate quantitative method.
	10 IR absorption spectroscopy	Use to establish structure and identity of molecules and quantitative analysis.
	11 Atomic emission spectroscopy	Highly sensitive identification of metals, limited sensitivity for halogens and non-metals.
	12 Atomic absorption spec.	Fast, reliable quantitative analysis of a given metal.
	13 Fluorescence spectroscopy	Highly sensitive, highly selective method for trace compounds or elements.
Mass	14 Mass spec.	Sorts ions by their mass–charge ratio, high sensitivity and high accuracy.
Thermal	15 Differential thermal analysis	Determines temperature of transitions and reactions, used for thermal stability and phase diagrams.
	16 Thermo-gravimetry	Determines weight as a function of temperature, used for thermal stability and compositional analysis.

detected by their colour, or spraying the paper with a reagent that produces a visible colour or glows under ultraviolet light. The distance moved by the spot is used to identify the component while its colour intensity is a measure of its concentration.

2 Thin layer chromatography

This method is the same as that used in paper chromatography (see item 1) with the paper being replaced by a thin layer of adsorbing substance such as silica gel coated onto a glass or plastic plate. This method tends to be more sensitive than paper chromatography.

Further reading: Fried, B. and Sherma, J. (1982), *Thin Layer Chromatography Techniques and Applications*, Marcel Dekker.

3 Liquid chromatography

Pumps are used to move a liquid up a column packed with a material such as ion exchange resins (Figure 11.2). The sample is injected into the base of the column and the times taken for the different constituents to reach the top of the column monitored by means of a detector. The most commonly used detector is the *ultraviolet absorption detector*. There are three basic forms of this detector, a fixed wavelength detector where just a few UV wavelengths can be selected by means of filters and the absorption of these by the sample is measured, a variable wavelength detector where a wide selection of wavelengths is possible and the absorption at each is scanned in turn (this being essentially a spectrophotometer), and one where the wavelength spectrum is simultaneously monitored over a range of wavelengths using an array of solid state diodes.

Another detector is the *refractive index detector* which involves monitoring the difference in refractive index between the mobile phase and the liquid emerging from the column. Liquid chromatography gives better reproducibility, resolution and accuracy than thin layer chromatography but is generally less sensitive. Unlike gas chromatography this method is not limited by the sample volatility or thermal stability and is used with organic and ionic compounds.

Further reading: Snyder, L. R. and Kirkland, J. J. (1974), *Introduction to Modern Liquid Chromatography*, Wiley; Willard, H. H., Meritt, L. L., Dean, J. A. and Settle, F. A. (1988), *Instrumental Methods of Analysis*, Wadsworth.

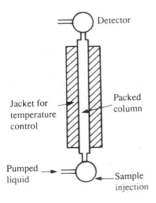

Figure 11.2 Liquid chromatography

4 Gas chromatography

Gas chromatography is a widely used on-stream analyser. A sample of the test vapour or gas is introduced into a stream of carrier gas, e.g. argon, helium or nitrogen, which then moves over a stationary phase (Figure 11.3). It is thus restricted to gaseous samples or samples that are volatile enough to be vaporized without decomposing. In *gas–liquid chromatography* the stationary phase is a thin layer of non-volatile liquid coated onto solid particles in a column (the solid particles act only as a support for the liquid). In *gas–solid chromatography* the solid particles provide the stationary phase. The packed column through which the moving phase moves is a long coiled tube. The rates at which the components of the sample pass through the column depend on their relative interactions with the stationary liquid or solid. The gas output from the column is monitored by means of a suitable detector which gives an output as a series of peaks spaced in time, each peak related to a particular component of the sample.

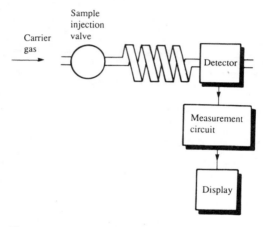

Figure 11.3 Gas chromatograph

The most commonly used detectors are the *katharometer* (Figure 11.4) and the *flame ionization detector* (Figure 11.5). The katharometer is based on the thermal conductivity of a gas depending on its composition (see Chapter 8, item 2). Thus the temperature and hence the resistance of an electrically heated wire is affected by the gas composition. Four such heated wires are used with two having just the carrier gas passing over them and the other two the gas emerging from the column. The four wires are connected as the four arms of a Wheatstone bridge and so the out-of-balance voltage gives an output related to the composition of the gas. The katharometer has a detection sensitivity from about 1 ng to 100 000 ng. It responds only to a restricted range of compounds, in general those just containing nitrogen or phosphorus. With the flame ionization detector, the gas emerging from the column passes into a hydrogen–oxygen flame. This causes the molecules in the gas to ionize and hence produce an ionization current between a pair of electrodes, between which there is a potential difference. The ionization current, after amplification, gives an output related to the number of $-CH_2-$ groups in the component

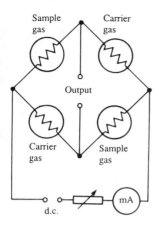

Figure 11.4 Katharometer

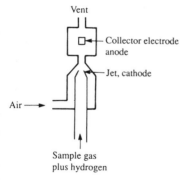

Figure 11.5 Flame ionization detector

entering the flame at the instant concerned. Thus for compounds containing this group the detector can be considered to be mass sensitive rather than concentration sensitive. The flame ionization detector has a detection range from about 0.01 ng to 10 000 ng. Only compounds which produce charged ions when burnt in a flame can be detected, this however includes most organic compounds.

Further reading: Noltingk, B. E. (ed.) (1985), *Jones' Instrument Technology*, vol. 2 (*Measurement of Temperature and Chemical Composition* (4th edn), Butterworth-Heinemann; Grob, R. L. (1977), *Modern Practice of Gas Chromatography*, Wiley; Willard, H. H., Meritt, L. L., Dean, J. A. and Settle, F. A. (1988), *Instrumental Methods of Analysis*, Wadsworth.

Electrochemical analysis

5 Electrical conductivity

The conductivity of an electrolyte depends on the ions present in the solution and their concentration. For concentrations less than about 10^{-4} mol/litre

$$\text{conductivity} = znc(\lambda^0{}_+ + \lambda^0{}_-)$$

where z is the charge on an ion, n the number of these ions produced by the dissociation of one molecule of the salt, c the concentration in mol/litre, $\lambda^0{}_+$ the positive ion conductivity at infinite dilution and $\lambda^0{}_-$ the negative ion conductivity at infinity dilution. These ionic conductivities depend on the ions concerned and are temperature dependent.

Conductivity, the reciprocal of resistivity, is defined as

$$\text{conductivity} = \frac{L}{RA}$$

where L is the length of the sample, A its cross-sectional area and R its electrical resistance. The unit of conductivity is $\Omega^{-1}m^{-1}$, however the unit S (siemen) is usually used for Ω^{-1} and it is usually to express the length and area in centimetres rather than metres, hence the unit S/cm.

A conductivity measurement involves the measurement of the electrical resistance between two electrodes immersed in the liquid a fixed distance apart. For a particular cell the conductivity is related to the resistance R by

$$\text{conductivity} = a/R$$

where a is a constant for the cell. Cells are available with cell constants varying from $0.01\ cm^{-1}$ to $100\ cm^{-1}$. A low cell constant is required for low conductivity measurements, a high one for high conductivities. The cell constant range enables measurements to be made from $0.05\ \mu S/cm$ to $200\,000\ \mu S/cm$. The cell constant is determined by measuring the resistance when the cell is filled with a liquid of known conductivity, a standard solution of potassium chloride being used.

The technique used for the measurement of the cell resistance is either by means of a Wheatstone bridge or direct measurement of the current through the cell for a specific potential difference across it. In both cases alternating current is used. This is because d.c. can lead to polarization problems. With both methods temperature compensation is necessary and a number of methods are available for this. For example, with the Wheatstone bridge a thermistor might be included in an arm adjacent to the cell, the change in resistance with temperature of the thermistor cancelling out the effect of temperature on the cell. An example of the use of this method is the measurement of the concentration of sulphur dioxide in air, this being achieved by measuring the conductivity of a reagent before and after it has absorbed the sulphur dioxide.

Further reading: Willard, H. H., Meritt, L. L., Dean, J. A. and Settle, F. A. (1988), *Instrumental Methods of Analysis*, Wadsworth.

6 Galvanic cell

When a metal electrode is placed in an electrolyte positive ions leave the metal and enter the electrolyte, making the electrode negatively charged, and positive metallic ions in the electrolyte may become deposited on the electrode and so give it a positive charge. The equilibrium value of the electrode potential with respect to the solution depends on the metal used and the concentration of metallic ions in the electrolyte. When two electrodes are in an electrolyte the e.m.f. developed between the two is the potential of one electrode with respect to the other. In order to obtain the electrode potential with respect to the solution for one electrode a standard electrode, having a known electrode potential with respect to the electrolyte, is used for the other electrode. Standard electrodes that are commonly used are the silver–silver chloride electrode and the mercury–mercurous chloride or calomel electrode (see Figure 8.21).

For the measurement of ion concentration in an electrolyte a high impedance voltage measurement circuit is used to measure the potential difference occurring between an ion-selective electrode (see Figure 8.21) and a standard electrode. Ion-selective electrodes have been developed for many ions, e.g. hydrogen, sodium, potassium, ammonium, silver, copper, lead, cadmium, chlorine, bromine, iodine, etc. The measurement of hydrogen ion concentration gives the pH value for the solution (see Chapter 8).

Further reading: Noltingk, B. E. (ed.) (1985), *Jones' Instrument Technology*, vol. 2 (*Measurement of Temperature and Chemical Composition*), (4th edn), Butterworth-Heinemann; Bailey, P. L. (1976), *Analysis with Ion-selective Electrodes*, Heyden; Willard, H. H., Meritt, L. L., Dean, J. A. and Settle, F. A. (1988), *Instrumental Methods of Analysis*, Wadsworth.

7 DC Polarography

The term polarography is used for *d.c. voltammetry* when a dropping mercury electrode is used. *Voltammetry* is essentially the measurement of the voltage–current relations for a cell containing an electrolyte. Figure 11.6 shows the basic form of the d.c. polarography instrument. A potential difference is applied between two electrodes, the anode being the pool of mercury in the bottom of the cell and the cathode being the reservoir of mercury which is connected to a fine capillary

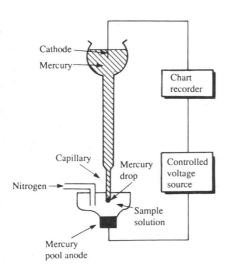

Figure 11.6 DC polarography

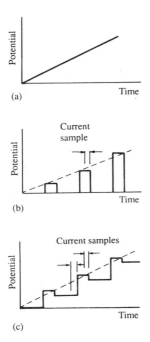

Figure 11.7 (a) Linear potential sweep (b) normal pulse (c) differential pulse voltametry

tube from which drops of mercury fall at the rate of about one every 3 or 4 s through the sample solution. Nitrogen is usually passed through the sample to purge it of dissolved oxygen. The voltage across the electrodes is increased at a controlled rate (Figure 11.7(a)) and the current in the circuit monitored. The size of the current is a measure of specific ions in the sample solution.

The above represents just the basic principle of d.c. polarography, further refinements are used to accentuate the changes in the output current. Thus in *current sampled voltammetry* the current is sampled at a fixed time after the birth of each mercury drop. In *normal pulse voltammetry* the voltage between the electrodes is not increased in a steady manner but held at a constant value until the drop is within about 60 ms of falling off and then abruptly increased for the remainder of the drop life before being reduced again for the next drop (Figure 11.7(b)). The square-wave voltage pulse is repeated for successive drops but with steadily increasing values. The current in each case is sampled during the last 20 ms of the life of each drop. In *differential pulse voltammetry* square-wave pulses, of constant size, are superimposed on a steadily increasing d.c. voltage (Figure 11.7(c)). The current is sampled prior to a pulse and during the latter stages of the pulse and just before the drop falls. The difference between these two current samples, for each pulse, is then plotted against the d.c. voltage. Polarography is used for the determination of the amounts of bismuth, copper, cobalt, nickel, lead, tin and zinc in light alloys, and trace and toxic elements in foodstuffs and pharmaceutical products.

Further reading: Heydrovsky, J. and Zuman, P. (1968), *Practical Polarography*, Academic Press; Willard, H. H., Meritt, L. L., Dean, J. A. and Settle, F. A. (1988), *Instrumental Methods of Analysis*, Wadsworth.

8 Anode stripping voltammetry

Anode stripping voltammetry uses essentially the same form of apparatus as polarography. A potential difference is applied between the electrodes for a period of time, often from 5 to 30 minutes, and a deposit produced on one of the electrodes. Then the process is reversed and a gradually increasing potential applied to strip the deposit off the electrode (as in Figure 11.7(a)) and the current recorded as a function of the potential. With anode stripping voltammetry metals such as bismuth, copper, lead, cadmium and zinc are deposited in the first stage of the procedure on the electrode that becomes the anode for the linear sweep potential stage.

The above is a description of the basic technique. Improvements are produced, in the same way as with polarography, by using differential pulse voltammetry. The method is more sensitive than polarography and can be used for analysis down to about 10^{-11} M solutions or about 0.1 μg per litre.

Spectroscopy

Spectroscopy is concerned with the measurement and identification of electromagnetic radiation that has been emitted, scattered or absorbed by atoms, molecules or other grouping of atoms.

9 UV and visible absorption spectroscopy

Instruments for detecting and measuring the intensity of the wavelengths absorbed by molecules in the ultraviolet and visible parts of the electromagnetic spectrum consist of a UV/visible light source emitting a continuous spread of wavelengths, a wavelength selector which selects the wavelength to be passed through the sample and then a photodetector to detect the intensity of light transmitted and give a read-out on some display. The source for the ultraviolet is generally a hydrogen or deuterium discharge lamp, while for the visible an incandescent filament lamp is used. Wavelength selection to isolate just a narrow band of wavelengths can be by filters (absorption or interference), a diffraction grating or a prism. Detectors can be photodiodes, photoemissive cells or photomultiplier tubes. With photodiodes, because they can be made very small, a linear or two-dimensional array may be used.

The simplest form of instrument has just a single beam of radiation (Figure 11.8(a)). The absorption at a particular wavelength for the sample is compared with the absorption for a reference standard, e.g. the sample in a solvent in a cell compared with just the solvent in a cell, as a result of two separate measurements. In a double beam instrument (Figure 11.8(b)) the radiation is split into two beams, one beam passing through the sample and the other through the reference standard. The radiation is usually chopped so that alternate pulses of radiation pass through the sample and the reference and hence when the signals are recombined the output alternates between the two. This enables the absorption comparisons to be made while the wavelength is continuously varied and so the output on a recorder can be a plot of absorption against wavelength.

Absorption in the ultraviolet and visible regions is a highly sensitive

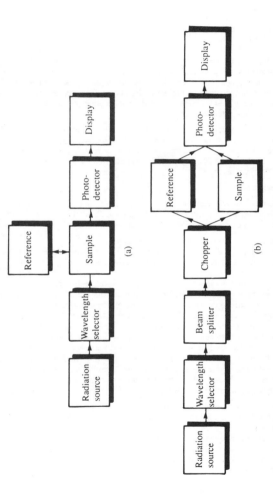

Figure 11.8 Form of (a) single beam (b) double beam spectrophotometers

and accurate means of quantitative analysis, it is however not so useful as other spectroscopic methods for identifying compounds.

Further reading: Willard, H. H., Meritt, L. L., Dean, J. A. and Settle, F. A. (1988), *Instrumental Methods of Analysis*, Wadsworth.

10 Infrared absorption spectroscopy

Infrared absorption spectroscopy gives information about the twisting, bending, rotating and vibrational motions of atoms in molecules and is used to determine the structure and identity of organic and inorganic compounds by identification of the characteristic absorption spectra of functional groups in the molecules. It is used for both identification and quantitative analysis.

There are two types of infrared instrument, dispersive and non-dispersive. The dispersive instrument has a wavelength selector, a prism or a diffraction grating to disperse the radiation, and is of the same form as the double beam UV and visible spectrophotometer described in Figure 11.8(b). With the non-dispersive instrument filters are generally used to restrict the wavelength band examined but there is no scanning of the absorption by wavelength, just an examination of the absorption at particular wavelength regions. Figure 11.9 shows the basic form of such an instrument. The radiation from the source is split into two parts, one passing through the sample and the other through the reference standard. The two beams are then combined by means of a chopper so that the output alternates from sample to reference. A filter is then used to select the wavelength region which is monitored by the detector and used to give a display. Such instruments are used where repetitive measurements are made of a known compound.

Different radiation sources, optical systems and detectors are required for different parts of the infrared spectrum. Thus for dispersive instrument for the near-infrared, $0.8\,\mu m$ to $2.5\,\mu m$, the source of radiation is a tungsten filament lamp, the optical system uses quartz prisms or reflection diffraction gratings, and the detector a photoconductive cell. For the mid-infrared, $2.5\,\mu m$ to $50\,\mu m$, a Nernst glower (a rod made from a fused mixture of oxides), a Globar (a silicon carbide rod) or incandescent coil of nichrome wire is used for the radiation source, the optical system is either a number of diffraction gratings with either a quartz prism or a filter, and the detector is a thermopile, a pyroelectric crystal (such crystals have surface charges which depend on the crystal temperature) or thermistor. For the far-infrared, $50\,\mu m$ to $1000\,\mu m$, a high pressure mercury-arc lamp is used for the radiation source, a reflection grating for the lower part of the range and an interferometer arrangement for the higher part, and the detector is a Golay pneumatic detector (the radiation causes gas in a cell to increase in pressure and so deform a diaphragm) or a pyroelectric crystal.

Further reading: Willard, H. H., Meritt, L. L., Dean, J. A. and Settle, F. A. (1988), *Instrumental Methods of Analysis*, Wadsworth.

11 Atomic emission spectroscopy

This is concerned with the radiation emissions from the excited electronic states of atoms. The pattern of wavelengths emitted is characteristic of the atom concerned and acts as a unique fingerprint in its identification. It is a highly sensitive identification method for all metallic elements with limited sensitivity for halogens and other non-metals.

Combustion in a flame provides a means of both converting a sample into atoms and exciting them, for this reason the method is often

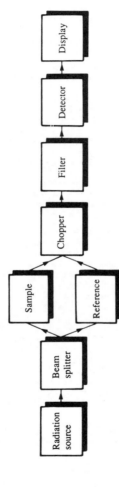

Figure 11.9 Form of a non-dispersive infrared spectrometer

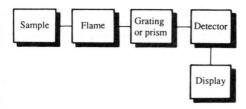

Figure 11.10 Form of an atomic emission spectrometer

referred to as *flame emission spectroscopy*. The sample, in solution, is introduced into a high velocity gas jet and breaks up into a fine aerosol. The resulting fine mist is then combined with the oxidiser–fuel mixture and burnt in the burner. The resulting radiation is then dispersed into its constituent wavelengths by a prism or grating and the individual wavelengths detected by a photodetector and the output displayed or recorded on a photographic plate (Figure 11.10).

Further reading: Willard, H. H., Meritt, L. L., Dean, J. A. and Settle, F. A. (1988), *Instrumental Methods of Analysis*, Wadsworth.

12 Atomic absorption spectroscopy

Atomic absorption spectroscopy is used for the quantitative analysis of elements in a sample. Radiation at the first resonance line (the transition between the ground state and the first excited state) is produced by a hollow cathode lamp and passed through the sample in its atomic state, combustion of the sample in a flame provides the means of converting the sample into its constituent atoms. A monochromator, typically a diffraction grating, is then used to isolate just the radiation emerging from the sample at the resonance line. The intensity of the wavelength is measured with a photodetector and the output displayed (Figure 11.11). The method gives a fast, reliable quantitative analysis for a given element, usually a metal.

Further reading: Willard, H. H., Meritt, L. L., Dean, J. A. and Settle, F. A. (1988), *Instrumental Methods of Analysis*, Wadsworth.

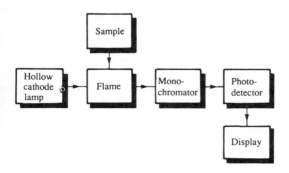

Figure 11.11 Form of an atomic absorption spectrophotometer

13 Fluorescence spectroscopy

Radiation incident on molecules and atoms can cause them to become excited and absorb radiation. Fluorescence occurs when these atoms or molecules return to their ground state by the emission of radiation at the incident wavelength or different ones. In general a *fluorometer* consists of a source of radiation (often a high pressure d.c. xenon lamp or low-pressure mercury vapour lamp), a filter or diffraction grating to isolate a specific wavelength of this radiation which is then incident on a cell containing the sample. The resulting fluorescence is generally detected in a direction at 90° to the incident radiation after passing through another filter or monochromator (Figure 11.12). With atomic fluorescence the sample cell is a flame in which combustion of the sample is used to obtain the sample in atomic form. Fluorescence spectroscopy is a highly sensitive, highly specific method of determining trace amounts of compounds or elements.

Further reading: Willard, H. H., Meritt, L. L., Dean, J. A. and Settle, F. A. (1988), *Instrumental Methods of Analysis*, Wadsworth.

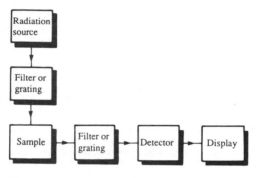

Figure 11.12 Form of a fluorometer

Mass spectrometry

14 Mass spectrometry

Mass spectrometry sorts the parent and fragment ions of molecules according to their mass-to-charge ratio and so enables both their identification and quantity to be determined. It is a highly sensitive method with high precision.

There are many different types of mass spectrometer. A common form of instrument consists of an inlet system by which the sample can be introduced into the instrument in which the pressure is about 10^{-5} Pa, a means of ionizing the sample, a mass analyser where electric and magnetic fields are used to separate the ions according to their mass-to-charge ratio, and an ion collection system which then records the ions according to their mass-to-charge ratio (Figure 11.13). The inlet system for liquids is often just injection by means of a hyperdermic needle through a rubber septum into a low pressure chamber where the liquid vaporizes and is then leaked through a small hole into the ionization chamber. Solids may be introduced into the inlet system on a probe and vaporized by heating. The ionization is commonly carried out by bombarding the vaporized sample with electrons from an

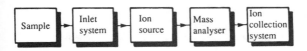

Figure 11.13 Form of a mass spectrometer

electron gun, the resulting positive ions being accelerated into the chamber by a suitably directed electric field. There are a number of forms of mass analysers. One form involves two elements, the first element being an electric field at right angles to the beam of ions in order to select ions which have the same kinetic energy, and then a magnetic field to select from those ions those with particular mass–charge ratios. Figure 11.14 shows such an arrangement in what is termed the *Matthaus–Herzog geometry*. The most commonly used ion collections system is an electron multiplier, the ions hitting a metal plate and causing electrons to be emitted which are then attracted to a second electrode where further electrons are emitted and then to a third electrode and so on through perhaps 15 to 18 electrodes with the number of electrons multiplying at each electrode. By varying the magnetic field different mass–charge ratio ions can be detected by the ion collection system. An alternative is to produce the complete mass spectrum on a photographic plate.

Further reading: Willard, H. H., Meritt, L. L., Dean, J. A. and Settle, F. A. (1988), *Instrumental Methods of Analysis*, Wadsworth.

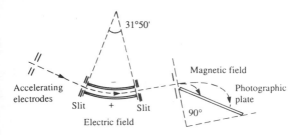

Figure 11.14 Mass spectrometer with the Matthaus–Herzog geometry

Thermal techniques

These techniques involve heating or cooling a sample and measuring some physical property of the material as a function of temperature.

Further reading: Wendlandt, W. W. (1986), *Thermal Methods of Analysis*, Wiley; Daniels, T. (1973), *Thermal Analysis*, Kogan Page.

15 Differential thermal analysis

Differential thermal analysis (DTA) involves heating the sample and a thermally inert reference material in two separate chambers in a common heating block at the same uniform rate while monitoring the difference in temperatures between them with a thermocouple, one junction being in the sample and the other junction in the reference material (Figure 11.15). In addition the temperature of the heating

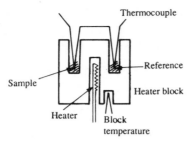

Figure 11.15 Differential thermal analysis

block is monitored. When a chemical or physical change occurs in the sample that results in the emission or absorption of heat its temperature changes from that in the reference material. A plot of the difference in temperature between the sample and reference materials against the temperature reveals such changes. The method is used to determine the thermal stability of materials and phase diagrams.

16 Thermogravimetric analysis

Thermogravimetric analysis (TGA) involves a sample being heated at a controlled rate in an inert or reactive atmosphere while its weight is continuously monitored. Changes in weight occur as a result of volatile products being formed or reaction products being produced. The results can be displayed as a graph of sample weight plotted against sample temperature. The method is used to determine the thermal stability of materials and compositional analysis.

12 Density

Density is defined as the mass per unit volume. *Relative density* is the ratio of the mass of a volume to the mass of an equal volume of water, usually at 4 °C. Table 12.1 lists the methods discussed in this chapter for the measurement of density and their main characteristics.

Further reading: Noltingk, B. E. (ed.), (1985), *Jones' Instrument Technology*, Butterworth-Heinemann.

Weight methods

1 Force-balance systems
Force–balance systems can be used for the continuous monitoring of the density of fluids in motion. Figure 12.1 shows the basic principles. The fluid passes through a horizontal U-shaped tube which has flexible connectors so that it can pivot about them. The weight of the fluid in the U results in a force on the tube support. With a pneumatic force–balance system, the force causes the pressure to change in a flapper–nozzle system. The changing pressure then results in a changing force applied to the tube support by means of a bellows arrangement. The result is that, at balance when the U is horizontal, the pressure in the system is a measure of the weight of the fluid in the U tube. Since the volume of the U is constant then the pressure is a measure of the density of the fluid. It is used for densities up to 1600 kg/m³ and with slurries or fluids containing solid matter, provided the flow rate is high enough to avoid deposition of entrained solids.

2 Load cell
If the level of a liquid in a container is kept constant then any change in its weight will be due to a change in density. Load cells (see Chapter 8, item 21 and also the use of load cells for level measurement in Chapter 17) in the supports of the container can be used to give continuous responses related to this change in weight.

Table 12.1 Density measurement systems

Principle	System	Characteristics
Weight	1 Force–balance	Continuous, can be used with slurries, densities up to 1600 kg/m³.
	2 Load cell	Continuous.
Buoyancy	3 Hydrometer	Simple, cheap.
	4 Torque tube	Can be used for high temperature liquids, open and sealed tanks.
Pressure	5 Differential pressure	Can be used for corrosive liquids.
	6 Bubbler	Not suitable for closed vessels or liquids containing particles.
Vibrating element	7 Vibrating tube	High accuracy, up to 3000 kg/m³, for liquids and liquid–solid mixtures.
	8 Vibrating cylinder	For gases up to 400 kg/m³
Radiation	9 Gamma	Small span of about 30 kg/m³ to 100 kg/m³.

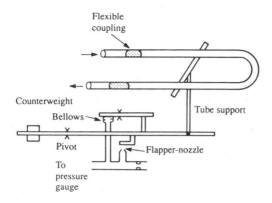

Figure 12.1 Force–balance density meter

Buoyancy methods

The upthrust acting on a body immersed or partially immersed in a fluid is equal to the weight of fluid it has displaced (Archimedes' principle).

3 Hydrometer

The simple hydrometer (Figure 12.2) is essentially a weighted tube which floats in a liquid to a depth which depends on the density of the liquid. The weight of the hydrometer is then equal to the weight of fluid displaced by the hydrometer. A scale on the stem of the instrument enables the density to be read off as the value at the liquid surface. The instrument is simple and cheap.

4 Torque tube

The torque tube (see Chapter 17 and Figure 17.9 where it is described for the determination of level) can be used for the measurement of density. A float is completely immersed and so, because the volume of liquid displaced does not change, the upthrust force on the float only changes if the density of the liquid changes. The upthrust force is used to twist a tube. This can be monitored by means of strain gauges or a pneumatic system. Such a method can be used for high temperature liquids and both open and closed tanks.

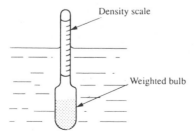

Figure 12.2 Hydrometer

Pressure methods

The difference in pressure between any two levels in a liquid or gas is equal to $h\rho g$, where h is the vertical distance between the levels, ρ the density and g the acceleration due to gravity.

5 *Differential pressure methods*

Figure 12.3 shows three methods that can be used for the measurement of the density of a liquid. In 12.3(a) the level of the liquid in the container is maintained at a constant level. Thus the pressure at some level below the surface depends only on the density. In 12.3(b) the pressure difference is measured between two different levels in the

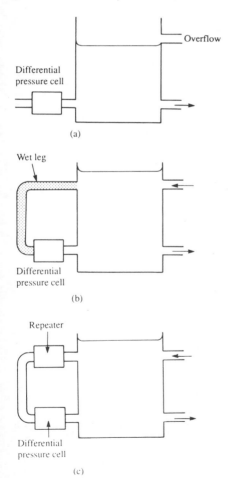

Figure 12.3 Differential pressure methods (a) constant level (b) with a wet leg (c) with a repeater

liquid, this being then proportional to the density of the liquid. Such a method is referred to as being with a *wet leg*. The wet leg generally contains a seal fluid which is denser than the process fluid and does not mix with it. Where a sealing liquid cannot be used a pressure repeater can be used (Figure 12.3(c)). The repeater reproduces the pressure at the upper level point at the lower level differential pressure instrument, so enabling the pressure difference between the two levels to be monitored.

6 Bubbler method

The bubbler method (see Chapter 17 and the use of this method for the measurement of level) is based on the principle that the pressure in an open tube in a liquid when fed with gas is limited by the gas escaping as bubbles. Two tubes are used with the same gas supply connected to both (Figure 12.4). The open ends of the tubes are at different depths in the liquid and so there is thus a difference in pressure produced between the two tubes. For a constant height difference between the tubes the pressure difference is related to the liquid density. Hence a measurement of the pressure difference is a measure of the density. Such a method is not suitable for closed vessels or liquids containing particles which might block the tubes. The method can be used for corrosive liquids since only the tubes dipping into the liquid need to be corrosion-resistant.

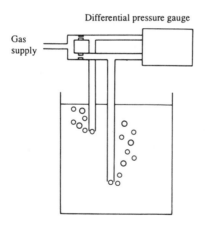

Figure 12.4 Bubbler method

Vibrating element methods

The frequency with which a wire or tube clamped at its ends will freely vibrate depends on the length between the clamps, the mass per unit length and the tension or stiffness (see Chapter 8, items 30 and 31).

7 Vibrating tube

Figure 12.5 shows the basic form of the vibrating tube method for the measurement of fluids flowing through a tube. The tube element is fixed at each end to heavy masses, so effectively clamping the ends. The tube is set into oscillation by means of magnetic forces supplied by alternating current in the drive coil, the coil being located at the

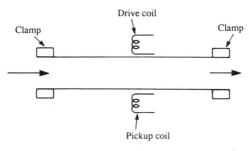

Figure 12.5 Vibrating tube

midpoint of the clamped length. The amplitude of the oscillation of the tube at the midpoint is monitored by a pick-up coil. The output from this coil is used as the feedback loop to the amplifier driving the drive coil. The result is that the tube is maintained in oscillation at its natural frequency. The frequency depends on the total mass of the tube and its contents. Thus since the tube has a constant volume the frequency is affected by changes in the density of the fluid in the tube. The relationship between the frequency f and the density ρ is of the form

$$\rho = \frac{A}{f^2} + \frac{B}{f} + C$$

where A, B and C are constants for the tube. This method can be used with liquids and liquid–solid mixtures, for densities up to 3000 kg/m^3 with high accuracy (about ± 0.2 kg/m^3).

8 Vibrating cylinder

For the measurement of the density of gases a thin walled cylinder is immersed in the gas. The cylinder is clamped at one end. The cylinder is set into vibration by means of an electromagnet drive coil and the amplitude of vibration monitored by means of a pick-up coil. The output from the pick-up coil is used as the feedback loop to the amplifier driving the drive coil and so maintains the oscillation at the natural frequency. Not only is the cylinder set into vibration but so also is the gas in contact with the cylinder. The density of this gas thus affects the effective mass per unit length of the cylinder and so the frequency. The relationship between the frequency and the gas density ρ is of the form

$$\rho = 2d_0\left(\frac{f_0 - f}{f}\right)\left[1 + \frac{K}{2}\left(\frac{f_0 - f}{f}\right)\right]$$

where f is the frequency with gas of density ρ, f_0 the frequency with a vacuum, and d_0 and K are constants for the tube. This method is used for densities up to 400 kg/^{3}.

Radiation methods

The transmission of gamma radiation through a material follows an exponential law of the form

$$I_t = I_i e^{-\mu\rho x}$$

where I_i is the incident radiation intensity, I_t the transmitted intensity after passing through a layer of material thickness x, density ρ and mass absorption coefficient μ. The mass absorption coefficient depends on the energy of the gamma radiation and the absorbing material concerned.

9 Gamma radiation density gauge

For a particular gamma source with the radiation passing through a constant thickness of the absorbing material, the fraction transmitted depends on the density. This is provided the changes in the material are not such as to significantly change the mass absorption coefficient. Typically a density gauge would consist of a radioactive source on one side of the containing vessel and an ion chamber detector (see Chapter 19) on the other. The output from the ion chamber is related to the density of the material in the vessel. Because the gauge measures the total density in the radiation path, problems can arise if deposits of entrained solids or bubbles occur between the source and detector. For radiation sources a typical size is 200 to 500 mC and this means an optimum path length between source and detector of about 0.3 m. The span of such an instrument is narrow, typically about 30 to 100 kg/m^3.

13 Displacement

This chapter includes the measurement of both linear displacement and angular displacement. Measurements of small displacements as part of the measurement of strain are included in Chapter 20. The term *comparator* is used for those instruments used in production engineering for the comparison of the length of an item with a length standard. Table 13.1 lists the methods discussed in this chapter and their main characteristics.

Table 13.1 Displacement measurement methods

Principle	System	Characteristics
Linear displacement		
Mechanical	1 Steel rule	Range 0 to 1000 mm, accuracy $\pm$0.1 to 0.3 mm.
	2 Micrometer	Range 0 to 600 mm, accuracy $\pm$0.002 mm.
	3 Callipers	Range 0 to 1000 mm, accuracy $\pm$0.02 to 0.06 mm.
	4 Dial gauge	Accuracy $\pm$0.005 to 0.02 mm, used for small displacements, pointer moving across a scale as output.
	5 Sigma comparator	Pointer moving across a scale as output.
	6 Johansson comparator	Pointer moving across a scale as output.
Pneumatic	7 Air gauge	Rotameter float movement as measure of displacement, can be used with different gauge heads for a range of measurements.
Electrical	8 Capacitor	High impedance, good resolution, accuracy $\pm$0.01%.
	9 Variable reluctance	Short range 0 to 10 mm, poor linearity, accuracy $\pm$0.5%.
	10 LVDT	Ranges vary from 0 to 2 μm to 0 to 500 mm, linear, accuracy $\pm$0.5%, robust, high reliability.
	11 Inductosyn	Used for machine tool control, accuracy $\pm$2.5 μm.
Optical	12 Split photocell	Range of only a few mm, discrimination about 1 μm, good stability.
	13 Moiré fringes	Used for machine tool control, reliable, discrimination about 1 μm.

continued

Table 13.1 (*continued*)

Principle	System	Characteristics
	14 Interferometer	Highly accurate, a few parts in a million, range up to 2 m.
	15 Time of flight	For measurements of large distances.
Angular displacement		
Electrical	16 Potentiometer	Angular displacements of 1 or many turns, accuracy $\pm 1\%$ to $\pm 0.002\%$.
	17 Resolver	Measures angular displacements of a shaft, resolution $0.4°$.
	18 Synchro	Can be used to measure angular displacement or as a pair to transmit angular displacements over distance.
Encoders	19 Incremental	Measures relative angular displacement, resolution typically $6°$ to $0.3°$.
	20 Coded	Measures absolute angular displacement, resolution typically $0.4°$ to $0.2°$.
Optical	21 Autocollimator	Measures small angular displacements, resolution 0.5 seconds.
	22 Angle Dekkor	Measures small angular displacements, resolution 0.2 minutes.

Mechanical measurements of linear displacement

Further reading: Brooker, K. (ed.) (1984), *Manual of British Standards in Engineering Metrology*, Hutchinson.

1 Steel rules

Engineers' steel measuring rules are available in lengths up to 1000 mm and have an accuracy of about ± 0.1 to 0.3 mm, the lower accuracy being for measurements involving the longer length rules.

2 Micrometer screw gauge

External micrometer screw gauges (Figure 13.1) can be used for the measurement of short lengths. They depend for their accuracy on the accuracy of the screw thread responsible for the movement of the anvil. Modern versions include the use of an electrical transducer to convert the movement of the screw into a digital display. The range of such instruments lie within 0 to 600 mm and they have an accuracy of about ± 0.002 mm. They are not suitable for continuous measurement.

3 Vernier callipers

Vernier callipers (Figure 13.2) are essentially just steel rules with jaws to enable the sides of the object being measured to be located. They have ranges within 0 to 1000 mm and have an accuracy of ± 0.02 to 0.06 mm.

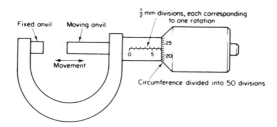

Figure 13.1 Micrometer screw gauge

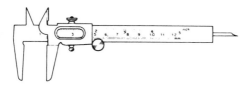

Figure 13.2 Vernier callipers

4 Dial gauge

The dial gauge (Figure 13.3) uses a system of gears to convert a linear displacement of its plunger into a highly magnified rotation of a pointer over a scale. With scale divisions of 0.01 mm, they are able to indicate small gradual changes of the order of ± 0.025 to within 0.003 mm, and have an accuracy of about ± 0.005 to 0.020 mm, the higher figure being for many revolutions of the pointer round the scale.

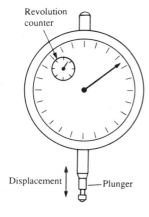

Figure 13.3 Dial gauge

5 Sigma comparator

The Sigma comparator (Figure 13.4) uses a compound lever to obtain large magnifications. The magnification of the first lever is y/x and that of the second lever R/r. Hence the overall magnification is yR/xr. Magnifications of the order of 300 to 5000 can be achieved.

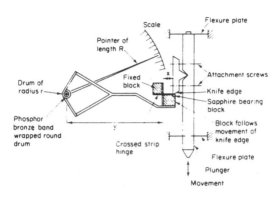

Figure 13.4 Sigma comparator

6 Johansson comparator

The Johansson comparator (Figure 13.5) uses a twisted metal strip to convert the movement of the plunger into a magnified rotation of a pointer over a scale. Magnifications as high as 5000 are possible.

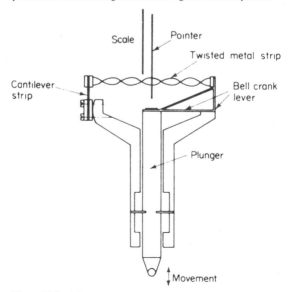

Figure 13.5 Johansson comparator

Pneumatic measurements of linear displacement

7 *Air gauge*

Air gauges, or pneumatic comparators, work on the principle that if an air jet is in close proximity to a surface, the rate of flow of air out of the jet depends on the distance of the surface from the jet. This flow rate is monitored using a rotameter flow gauge (see Chapter 15). Figure 13.6(a) shows the form such a gauge can take when used for linear displacements, Figure 13.6(b) when used for the measurement of hole diameters, Figure 13.6(c) for the measurement of hole taper, and

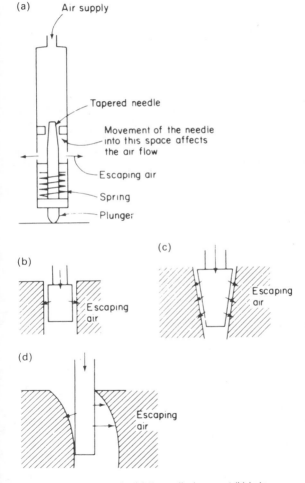

Figure 13.6 Air gauge for (a) linear displacement (b) hole diameter (c) hole taper (d) bore straightness

Figure 13.6(d) for bore straightness. Other forms of gauge are possible for other situations. The surface being gauged has to be smooth and non-porous. The movement of the float in the rotameter can be 5000 to 10 000 times more than the displacement actually being measured.

Further reading: Brooker, K. (ed.) (1984), *Manual of British Standards in Engineering Metrology*, Hutchinson.

Electrical measurements of linear displacement

There are many forms of electrical transducers that can be used to give responses related to displacement, e.g. the linear potentiometer (see Chapter 8 and Figure 8.4), capacitors (see Chapter 8 and Figures 8.6 and 8.8), variable reluctance (see Chapter 8 and Figure 8.10 or Figure 8.12), and the LVDT (see Chapter 8 and Figure 8.13).

8   *Capacitor*
Capacitive systems for displacement measurement have high output impedance (and as a consequence are very susceptible to noise), good resolution (can be of the order of nanometres), with accuracies up to about $\pm 0.01\%$.

9   *Variable reluctance*
Figure 13.7 shows one form of *variable reluctance comparator*. Movement of the plunger results in a displacement of the ferromagnetic strip between the two coils, increasing the reluctance of one and decreasing it for the other. Another form of comparator is shown in figure 8.12. In both cases the two coils are connected in adjacent arms of an a.c. bridge and the out-of-balance signal becomes a measure of the displacement. Such comparators have short ranges, typically 0 to 10 mm, poor linearity and an accuracy of about $\pm 0.5\%$.

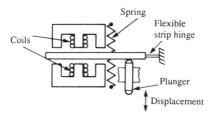

Figure 13.7   Variable reluctance displacement comparator

10   *LVDT*
The *linear variable differential transformer* (*LVDT*) is widely used for the measurement of displacements (see Chapter 8 for a full discussion). They are available with ranges which cover small displacements of the order of 0 to 0.2 µm or less to larger displacement of the order of 0 to 500 mm. They are robust, linear, have high reliability, high sensitivity, and an accuracy of about $\pm 0.5\%$.

11   *Linear inductosyn*
The linear inductosyn (Figure 13.8) consists of a track along which a slider moves, the slider being attached to the body whose position or displacement is to be measured. This might be a cutting tool since the linear inductosyn is widely used for the control of machine tools. The track, which may be several metres long, has a fine metal wire formed into a single, continuous, rectangular waveform. Typically the pitch of

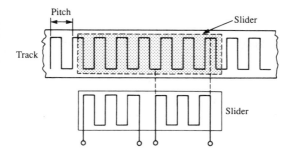

Figure 13.8 Linear inductosyn

the waveform is about 2 mm. The slider sits on top of the track and is much shorter, typically about 50 or 100 mm, and has two separate wires formed into the rectangular waveform. The waveforms are, however, displaced from each other by one-quarter of the cycle pitch. This means when one of the slider wire lengths is aligned with the track wire, the other slider wire is out-of-step by a quarter of a cycle (as in Figure 13.8). When an alternating current flows through the track wire e.m.f.s are induced in the slider wires. For one of the slider wires the relationship between the e.m.f. V_1 and the slider displacement x is

$$V_1 = kV \sin (2\pi x/p)$$

where k is a constant, V the value of the track voltage, and p the pitch of the waveform. For the other slider wire, since it is displaced by a quarter of a cycle, i.e. 90°, then its e.m.f. V_2 is related to the slider displacement by

$$V_2 = kV \sin [(2\pi x/p) + 90°] = kV \cos (2\pi x/p)$$

The sum of these two voltages is a voltage with an amplitude which is cyclic, repeating itself every time the displacement x changes by the pitch p. Because of this the inductosyn cannot unambiguously interpret displacements greater than the pitch, i.e. about 2 mm and is generally used in conjunction with another transducer which has coarser resolution but a larger range. The inductosyn has an accuracy of about $\pm 2.5\ \mu$m.

Optical measurements of linear displacement

12 Position sensitive photocells
With the *split cell* form of position sensitive photocell (Figure 13.9) a defined beam of light falls on a photocell which is split down the middle. When the beam is central then equal segments of each photocell are illuminated. A displacement of the light beam however results in more of one cell being illuminated than the other with the result that a differential amplifier gives an output. Such an instrument has a displacement range of a few millimetres, can detect changes of about 1 μm and has good stability.

13 Moiré fringes
Moiré fringes are produced when light passes through two gratings which have rulings inclined at a slight angle to each other. Figure 13.10(a) shows a transmission form of instrument and Figure 13.10(b) a reflection form. With both, the long grating is fixed to the object

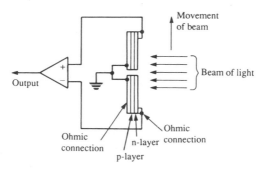

Figure 13.9 Split cell

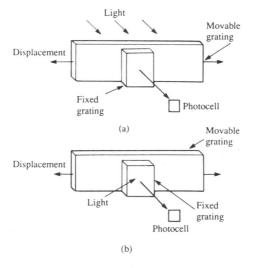

Figure 13.10 Moiré fringe method (a) transmission
(b) reflection

being displaced. With the transmission form each grating has a series
of parallel straight opaque lines between which are parallel clear lines
through which transmission of light occurs, while with the reflection
form the long grating has a series of parallel reflective lines and a short
grating with parallel transmission lines. Coarse gratings might have 10
to 40 lines per millimetre, fine gratings as many as 400 per millimetre.
Movement of the long grating relative to the fixed short grating results
in fringes moving across the view of the photocell and thus its output
oscillating up and down. Displacements as small as 1 μm can be
detected by this means. Such methods have high reliability and are
used for the control of machine tools.

14 Interferometer

Figure 13.11 shows the one form of the *laser interferometer* for the measurement of displacement. The helium–neon laser emits two light waves, differing in frequency by 2×10^6 Hz and plane polarized at right angles to each other. At the first beam splitter, each frequency is split into two portions with one portion serving as a reference and passing to photodetector A. There the two frequencies are combined to produce a beat frequency fluctuation in output of 2×10^6 Hz. The second beam splitter separates the two frequencies. The light at frequency f_1 passes to a fixed reflector, a corner cube, and is reflected back through the beam splitter to photodetector B. The light at frequency f_2 passes to the movable reflector, a corner cube, and is reflected back to the beam splitter and so to photodetector B. With the movable reflector stationary the output from photodetector B is the same as that from photodetector A. However, when A is moving there is a Doppler shift in frequency f_2 and so the output from B then differs from that of A. The outputs from A and B are subtracted. The result is an oscillating signal, the number of peaks of which is related to the displacement, and the frequency of which is related to the velocity. Such a method is highly accurate, a few parts in a million, and can be used over ranges up to about 2 m.

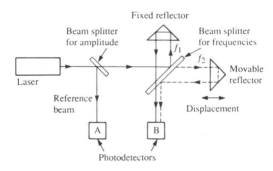

Figure 13.11 Interferometer

15 Time of flight

In time of flight methods of determining distances a pulse of radiation, generally from a laser, is emitted and the time measured for it to travel to a distant reflector and back. Because of the high speed of light, in air about 3×10^8 m/s, this method is only suitable for long distances when the time of flight is sufficiently large to be measured with reasonable accuracy.

Electrical angular displacement measurements

16 Circular and helical potentiometers

See item 4 Chapter 8 for a discussion of potentiometers. Circular and helical potentiometers give an output of a change in resistance for an angular rotation of the shaft moving the wiper over the resistance track. Circular potentiometers have a track of not more than one turn, i.e., no more than an angular displacement of $360°$, and an accuracy of about $\pm 1\%$. Helical potentiometers often have many turns and an accuracy which can be as good as $\pm 0.002\%$.

17 Resolver

Resolvers (Figure 13.12) have two stator windings, at right angles to each other, and a rotor. With the *varying amplitude output resolver* the stator is supplied with a single-phase sinusoidal voltage of frequency ω so that the amplitudes in the two stator windings are out of step by $90°$, i.e.

$$V_1 = V \sin \beta$$
$$V_2 = V \sin (\beta + 90°) = V \cos \beta$$

V is the voltage at some instant and is $V_s \sin \omega t$. The e.m.f. V_o induced in the rotor when it is at some angle θ is

$$V_o = kV \sin (\beta - \theta) = [KV_s \sin (\beta - \theta)] \sin \omega t$$

where k is a constant depending on the degree of magnetic coupling. The output thus has the same frequency as the input but has an amplitude which depends on the angle θ. A resolution of about $0.4°$ is possible.

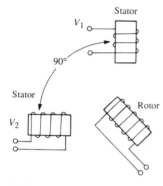

Figure 13.12 Resolver

18 Synchro

See item 13 Chapter 8 and Figure 8.15 for a discussion of synchros. A synchro differs from a resolver in having three stator windings positioned at $120°$ intervals round a rotor. A sinusoidal voltage is supplied to the rotor and the e.m.f.s induced in each of the stator coils depends on its orientation relative to the rotor position, i.e.

$$V_1 = kV \sin \theta$$
$$V_2 = kV \sin (\theta + 2\pi/3)$$
$$V_3 = kV \sin (\theta - 2\pi/3)$$

where V is the rotor voltage at an instant and is $V_r \sin \omega t$, k is a constant related to the degree of magnetic coupling and θ the angle of the rotor. The amplitude of each stator voltage is thus a measure of the angular position of the rotor. While synchros are used for the measurement of angular displacement a more common use is for a pair of them to be used to transmit angular displacement information over a distance (see item 13, Chapter 8 and Figure 8.16).

Shaft encoders

19 Incremental shaft encoder

See item 28, Chapter 8 and Figure 8.29 for a discussion of incremental shaft encoders. Such an encoder gives a digital output related to the angular position of a shaft relative to some datum position. The resolution is determined by the number of 'windows' that can be arranged on the shaft disc. Typically the number varies from 60 to over a thousand and so a resolution from 6" down to 0.3" or better is possible.

20 Coded shaft encoders

See item 29 Chapter 8 and Figure 8.30 for a discussion of coded shaft encoders. Such an encoder measures the absolute position of a shaft within one revolution. Typically encoders tend to have up to 10 or 12 tracks. A 10 track encoder gives 1024 counts per revolution and so a resolution of about 0.4°, a 12 track encoder with a count of 4096 a resolution of about 0.1°. Optional shaft encoder with as many as 21 tracks are possible, such an encoder giving a resolution of about one second.

Optical angular displacement measurements

See item 12 Chapter 9 and Figure 9.19 for a discussion of the principles involved.

21 Autocollimator

Figure 13.13 shows the basic principles of the autocollimator. An illuminated target wire is mounted at the focus of the collimating lens. The target wire and the reflected image of the target wire are viewed through an eyepiece which incorporates a scale graduated in 0.5 minute intervals. In addition there is a pair of setting wires, the position of which can be adjusted by means of a micrometer graduated in 0.5 second intervals. When the instrument is being used the setting wires are moved so that they straddle the target wire image and the reading taken. This is repeated after the reflector has rotated, the difference giving the angle rotation.

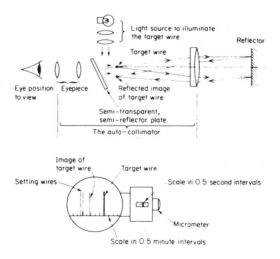

Figure 13.13 Autocollimator

22 Angle Dekkor

Figure 13.14 shows the basic principles of the Angle Dekkor. A scale engraved on a screen is illuminated, the screen being in the focal plane of the collimating lens. The light passing through the collimating lens is then reflected by the reflector back through the lens to be focused on the scale. The screen in the area where the reflected image is received has another scale, this being at right angles to the reflected image scale. The displacement of the reflected scale image above or below the datum line of the fixed screen scale gives a measure of the angle of the reflector in one particular direction. The movement of the reflected scale image along the fixed screen scale is a measure of the angle of the reflector in a direction at right angles to the first direction. The scale is graduated in minutes and it is possible to estimate to about 0.2 minute.

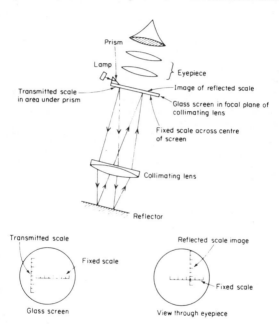

Figure 13.14 Angle Dekkor

14 Electrical quantities

Table 14.1 lists, by the quantity measured, the instruments and measurement methods discussed in this chapter.

Table 14.1 Electrical measurements

Measurement	Instrument	Characteristics
Current d.c.	1 Moving coil meter	F.S.D. from ¹0 µA to 20 mA, extended to 10 A with shunts, linear scale, accuracy up to ±0.1%.
	2 Light spot galvo.	Sensitivity 100 mm/µA, used often as null detector.
	7 Multimeter	F.S.D. typically 50 µA to 10 A, accuracy ±1%.
	10 Electronic inst.	F.S.D. 1 µA to 3 A, accuracy ±1 to 2%.
	13 Electronic multi.	F.S.D. 1 µA to 3 A, accuracy ±1%, 30 mV drop.
	14 Moving iron meter	F.S.D. from 0.1 A to 30 A, d.c. or a.c. up to about 100 Hz, non-linear scale, accuracy ±0.5%.
	17 Dynamometer	F.S.D. 5 mA to 20 A, a.c. or d.c., non-linear scale.
	26 Digital inst.	Accuracy ±0.2% + 1 digit, full scale 200 µA to 2 A.
Current a.c.	3 Moving coil meter	F.S.D. from 10 mA to 10 A with current transformer, bandwidth 50 Hz to 10 kHz, accuracy ±1%, scale only correct for sinusoidal waveform.
	7 Multimeter	F.S.D. typically 10 mA to 10 A, accuracy ±2%.
	11 Electronic inst.	F.S.D. 1 µA to 3 A, accuracy ±2 to 5%, bandwidth 20 Hz to 100 MHz.
	13 Electronic multi.	F.S.D. 1 µA to 3 A, accuracy ±2 to 5%, 30 mV drop, bandwidth 20 Hz to 100 MHz.
	14 Moving iron meter	F.S.D. from 0.1 A to 30 A, d.c. or a.c. to about 100 Hz, non-linear scale, accuracy ±0.5%.
	16 Thermocouple inst.	F.S.D. 2 mA to 50 mA, non-linear scale, bandwidth 10 Hz to 50 MHz, fragile.
	17 Dynamometer	F.S.D. 5 mA to 20 A, a.c. or d.c., non-linear scale.

continued

Table 14.1 (*continued*)

Measurement	Instrument	Characteristics
	26 Digital inst.	Accuracy $\pm 1\%$ + 2 digits, full scale 200 μA to 2 A, frequency 45 Hz to 1 kHz.
Voltage d.c.	4 Moving coil meter	F.S.D. from 50 mV to 100 V with multipliers, linear scale, accuracy up to $\pm 0.1\%$.
	7 Multimeter	F.S.D. typically 100 mV to 3 kV, accuracy $\pm 1\%$.
	8 Electronic inst.	Input impedance about 10 MΩ, f.s.d. 15 mV to 1000 V, accuracy $\pm 1\%$.
	13 Electronic multi.	Input resistance up to 100 MΩ, f.s.d. 10 mV to 1000 V, accuracy $\pm 1\%$.
	15 Moving iron meter	Low input impedance, about 50 Ω/V, minimum f.s.d. of 50 V, d.c. or a.c. up to 100 Hz.
	18 Dynamometer	Non-linear scale, d.c. or a.c. up to 2 kHz.
	23 Electrostatic inst.	High impedance, accurate, range 100 V to 1 kV, a.c. or d.c.
	24 Digital inst.	Accuracy at least $\pm 0.1\%$ + 1 digit, full scale 100 mV to 1000 V, input resistance 10 MΩ or higher.
	33 Potentiometer	Measures e.m.f. and can be adapted for thermoelectric e.m.f.s.
Voltage a.c.	5 Moving coil meter	F.S.D. from 3 V to 3000 V with transformer, linear scale, accuracy $\pm 1\%$, bandwidth 50 Hz to 10 kHz.
	7 Multimeter	F.S.D. typically 3 V to 3 kV, accuracy $\pm 2\%$.
	9 Electronic inst.	Input impedance about 10 MΩ, f.s.d. 100 μV to 1000 V, bandwidth 20 Hz to 100 MHz, accuracy ± 2 to 5%.
	13 Electronic multi.	Input capacitance 2 pF, impedance up to 10 MΩ, f.s.d. 10 mV to 1000 V, bandwidth 20 Hz to 100 MHz, accuracy ± 2 to 5%.
	15 Moving iron meter	Low input impedance, about 50 Ω/V, minimum f.s.d. of 50 V, d.c. or a.c. up to 100 Hz.

Table 14.1 (*continued*)

Measurement	Instrument	Characteristics
	16 Thermocouple inst.	10 Hz to 50 MHz, fragile, non-linear scale.
	18 Dynamometer	Non-linear scale, d.c. or a.c. up to 2 kHz.
	23 Electrostatic inst.	High impedance, accurate, range 100 V to 1 kV, a.c. or d.c.
	25 Digital inst.	Accuracy at least $\pm 1\% + 3$ digits, full range 100 mV to 1000 V, impedance 10 MΩ/100 pF, frequency 10 Hz to 10 kHz.
Resistance	6 Ohmmeter	Non-linear scale.
	7 Multimeter	Non-linear scale, f.s.d. typically 2 kΩ to 20 MΩ, accuracy $\pm 3\%$ of mid scale.
	12 Electronic inst.	Non-linear scale, ranges up to about 100 MΩ, accuracy $\pm 3\%$ of mid scale.
	13 Electronic multi.	Non-linear scale, ranges up to about 100 MΩ, accuracy $\pm 3\%$ of mid scale.
	27 Digital inst.	Accuracy at least $\pm 0.1\% + 1$ digit, full scale 200 Ω to 1000 MΩ.
	28 Wheatstone bridge	Range 1 Ω to 10 MΩ.
	29 Kelvin double br.	Range 0.1 $\mu\Omega$ to 1 Ω.
	30 High res. bridge	High resistance measurement.
	34 Single ratio bridge	Accurate, only a few standard components needed.
	35 Double ratio bridge	Accurate, only a few standard components needed.
	36 Q-meter	
Power	19 Dynamometer	Power up to 300 V $\times$ 20 A, minimum 5 W, d.c. or a.c. up to 400 Hz, accuracy ± 0.1 to 0.5%.
	21 Electronic inst.	Power 0.1 W to 100 kW, up to 100 kHz, accuracy $\pm 0.5\%$ to 1%.
Power factor	20 Dynamometer	
Q-factor	36 Q-meter	
Energy	22 Watthour meter	For measurement of electrical energy used from mains supply.

continued

Table 14.1 (continued)

Measurement	Instrument	Characteristics
Inductance	31 AC bridges	
	34 Single ratio bridge	Accurate, only a few standard components needed.
	35 Double ratio bridge	Accurate, only a few standard components needed.
	36 Q-meter	
Capacitance	32 AC bridges	
	34 Single ratio bridge	Accurate, only a few standard components needed.
	35 Double ratio bridge	Accurate, only a few standard components needed.
	36 Q-meter	
Frequency	37 Digital counter	Range 0 to 200 MHz, accuracy 1 in 10^5 to 10^9.
	38 Lissajous figures	Frequency ratio and phase measurements.
	32 Wien bridge	Accurate for audio frequencies.

Moving coil meter

See item 1, Chapter 10 and Figure 10.1 for details of the permanent magnet moving coil meter. Loading effects (see Chapter 5) have to be considered when such meters are used.

1 Current d.c.

The basic meter movement employing springs to provide the restoring torque and a pointer moving over a scale is likely to have a minimum d.c. full scale current of between $10\,\mu A$ and 20 mA. Higher d.c. current full scale deflections, up to typically about 10 A, can be obtained by the use of shunts, see item 15, Chapter 9 and Figure 9.25. Figure 14.1 shows the form of the universal shunt. The sensitivity, i.e. angular deflection of the coil per unit current, of a shunted meter is given by

$$\text{sensitivity} = \frac{R_s S}{R_s + R_g}$$

where R_g is the resistance of the galvanometer, i.e. that of the coil and any series resistor included with the coil, R_s the resistance of the shunt and S the sensitivity of the unshunted movement. Such instruments have linear scales and accuracies up to about $\pm 0.1\%$ of full scale deflection. They are generally unaffected by stray magnetic fields.

2 Light spot galvanometer

For use as a null detector in a d.c. bridge where very small currents need to be detected, a *light spot moving coil galvanometer* (Figure 14.2) can be used. This has a coil with a large number of turns suspended from a metal strip, the twisting of this strip providing the restoring torque. Rotation of the coil causes a beam of light to be reflected across a scale. Such an instrument is likely to have a current sensitivity of about 100 mm of movement of light spot across the scale per microamp.

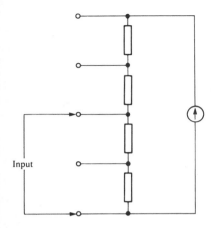

Figure 14.1 Universal shunt

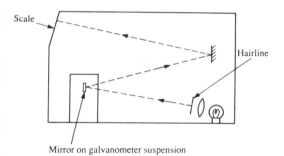

Figure 14.2 Light spot moving coil galvanometer

3 Current a.c.

A bridge rectifier, as in Figure 9.27, can be used to convert a.c. into a direct current for measurement by a moving coil meter. The meter responds to the average value of the current but the scale is calibrated in terms of the root-mean-square value assuming the current is sinusoidal. If this is not the case the scale is in error, the error being

$$\text{error} = \left(\frac{1.11 - F}{F} \right) \times 100\%$$

where F is the form factor. Different current ranges are obtained by using a current transformer with different tappings for the various ranges. Typically the a.c. full scale deflections vary from 10 mA to 10 A. The accuracy is up to $\pm 1\%$ of full scale deflection with a bandwidth of about 50 Hz to 10 kHz.

4 Voltmeter d.c.

The permanent magnet moving coil meter gives voltage readings as a result of measuring the current through a resistance, for the lowest

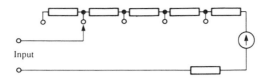

Figure 14.3 Multipliers

range this being just the resistance of the meter coil. Multipliers (see item 16, Chapter 9 and Figure 9.26) can be used to extend the range (Figure 14.3). The sensitivity of a meter with multiplier is given by

$$\text{sensitivity} = \frac{S}{R_m + R_g}$$

where R_g is the resistance of the galvanometer coil, R_m the resistance of the series multiplier and S the sensitivity of the meter without a multiplier. Typically full scale deflections vary from 50 mV to 100 V with an accuracy up to $\pm 0.1\%$ of full scale deflection and a linear scale. The resistance of the voltmeter R_m has to be considered in relation to that of the circuit R_{Th} to which it is connected for any loading effects (see Chapter 5), the accuracy being given by

$$\text{accuracy} = \frac{R_m}{R_m + R_{Th}} \times 100\%$$

Thus for a 99% accuracy R_m must be at least 99 times larger than R_{Th}. For an instrument to have a high input resistance it must have a high resistance multiplier in series with an instrument movement which gives a low current full scale deflection.

5 Voltmeter a.c.
For use as a voltmeter the moving coil meter has a bridge rectifier, as in Figure 9.27, with a resistor in series with the bridge and the alternating voltage. The minimum full scale deflection of such an instrument is likely to be about 3 V with a voltage transformer being used to give ranges up to a full scale deflection of about 3000 V. The instrument has a linear scale but is only calibrated to give correct root-mean-square values for sinusoidal alternating voltages, corrections have to be made for other forms (see item 3 above). Accuracy is typically no better than $\pm 1\%$ within a frequency band of 50 Hz to 10 kHz. This type of instrument tends to have a low input impedance and consequently presents loading problems (see item 4 above).

6 Ohmmeter
Figure 14.4(a) shows one form of a basic ohmmeter circuit. A battery, e.m.f. E, is connected in series with the meter resistance R_g and any resistance R connected into the circuit, the current I then being a measure of the resistance in the circuit.

$$I = \frac{E}{R_g + R}$$

The relationship between the current and the resistance R is non-linear and hence the ohmmeter circuit gives a non-linear resistance scale. The zero of the resistance scale corresponds to a full scale current reading. A variable resistor is generally included in the circuit to compensate for changes in the e.m.f. of the battery. With the terminals of the ohmmeter short-circuited, i.e. $R = 0$, this resistor is adjusted so that

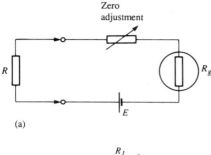

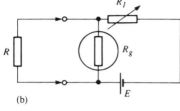

Figure 14.4 Ohmmeter circuits

the resistance reading is zero, i.e. the meter is indicating a full range current deflection.

Figure 14.4(b) shows another form of basic ohmmeter circuit, this form being particularly useful for low resistances. When R is infinity, i.e. open-circuit, and full-scale current obtained

$$I_{fsd} = \frac{E}{R_1 + R_g}$$

The unknown resistance is connected in parallel with the meter and acts as a shunt, so reducing the current passing through the meter.

$$\frac{E}{I} = R_1 + \frac{R_g R}{R_m + R}$$

The current through the meter I_g is thus given by

$$I = I_g + \frac{R_g I_g}{R}$$

Hence

$$I_g = \frac{ER}{R_1 R_g + R(R_1 + R_g)}$$

The meter reading thus depends on the value of the unknown resistance R.

7 Multimeter

A multimeter is a multi-range instrument using a permanent magnet moving coil instrument. A typical meter has full scale deflections for d.c. current ranges from 50 µA to 10 A, a.c. current 10 mA to 10 A, d.c. voltage 100 mV to 3000 V, a.c. voltage 3 V to 3000 V, resistance 2 kΩ to

20 MΩ, with d.c. accuracy of $\pm 1\%$ f.s.d., a.c. accuracy $\pm 2\%$ f.s.d. and resistance $\pm 3\%$ of the mid-scale reading.

Electronic instrument

The moving coil meter has the problems of low input impedance and low sensitivity on a.c. ranges. These can be overcome by using an amplifier between the input and the moving coil meter.

8 Voltage d.c.

Figure 14.5 shows a basic circuit for an electronic d.c. voltmeter. Such a meter would typically have an input impedance of about 10 MΩ, an accuracy of about $\pm 1\%$ and full scale deflections ranging from about 15 mV to 1000 V.

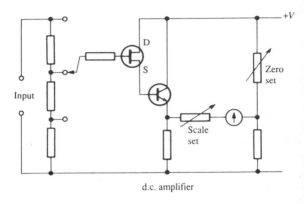

d.c. amplifier

Figure 14.5 Electronic d.c. voltmeter

9 Voltage a.c.

There are a number of versions of electronic voltmeter for alternating voltage determination. The mean or average-responding type gives an output which is a measure of the average value of the voltage waveform. Such a meter has rectification occurring either before or after the amplification (Figure 14.6). With such a meter the scale is usually calibrated in r.m.s. values assuming the input signal is sinusoidal. If this is not the case the scale is in error and corrections have to be made (see item 3). The peak-responding type gives an output which is a measure of the peak value of the voltage waveform. Such a meter uses a circuit in which the signal is half-wave rectified and then applied to a capacitor to charge it to the peak value of the voltage. This peak value is then amplified and used to give a deflection with a moving coil meter (Figure 14.7). The r.m.s.-responding type gives an output which is a measure of the r.m.s. value of the voltage waveform. Such an instrument may involve a thermocouple which gives an output related to the power generated by the amplified input voltage being applied to a resistor. The output from the thermocouple is then amplified by a d.c. amplifier and used to give a deflection which is a measure of the r.m.s. value of the input voltage (Figure 14.8). The effect of non-linearity in the response of the measuring thermocouple is cancelled by similar non-linearity in the response of another thermocouple in the feedback circuit of the amplifier. An alternative to using a thermocouple is to rectify the input voltage, use circuits to

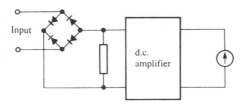

Figure 14.6 Average-responding electronic voltmeter

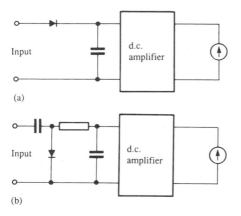

(a)

(b)

Figure 14.7 Peak-responding electronic voltmeter

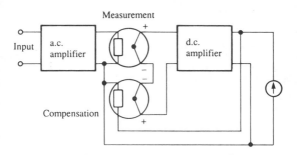

Figure 14.8 R.M.S.-responding electronic voltmeter

square the voltage, integrate it over a cycle and then take the square root and feed it to a moving coil meter. Typically these meters have full scale deflections from about $100\,\mu V$ to $1000\,V$, input impedance of about $10\,M\Omega$, a bandwidth of about $20\,Hz$ to $100\,MHz$, and an accuracy which varies from about ± 2 to 5% depending on the range and the frequency.

10 Current d.c.

For current measurement the amplifier plus meter can be regarded as a voltmeter and thus current is measured by determining the voltage drop across a known resistor. Typically ranges from full scale deflections of 1 μA to 3 A with an accuracy of about ± 1 to 2% of full scale reading and an input resistance which decreases from kΩ on the low current range to less than an ohm on the high current range.

11 Current a.c.

Current is measured by determining the voltage drop across a known resistor. Typically ranges extend from full scale deflections of 1 μA to 3 A with an accuracy of ± 2 to 5%, this depending on the range and the frequency. The bandwidth is about 20 Hz to 100 MHz.

12 Resistance

Unknown resistances are determined by measuring the voltage drop across them when supplied with a constant current. Different ranges are produced by using a voltage divider to present just a known fraction of the voltage drop to the electronic voltmeter. Typically such an arrangement is used to measure resistances, in a series of ranges, from 1 Ω to 100 MΩ. The scales are non-linear and have an accuracy of about ± 3% of the mid-scale value.

13 Multimeter

Electronic multimeters tend to have higher sensitivities than conventional multimeters, offering, through range and function selectors, measurements of a.c. and d.c. voltage, a.c. and d.c. current and resistance. Typically on the d.c. voltage range they have an input resistance of about 1 MΩ/V up to a maximum of 100 MΩ, on the a.c. voltage range an input capacitance of about 2 pF and impedance of 1 to 10 MΩ, and on a.c. and d.c. current ranges they produce about a 30 mV drop. Full scale deflections on the a.c. and d.c. voltage ranges typically vary from 10 mV to 1000 V, and on the a.c. and d.c. current ranges from 1 μA to 3 A. The accuracy of d.c. ranges is about ± 1% and a.c. ranges ± 2 to 5% depending on the range and the frequency. The bandwidth is about 20 Hz to 100 MHz. Resistances from about 1 Ω to 100 MΩ can be measured on a number of ranges with an accuracy of about ± 3% of the mid-scale value.

Moving iron meter

There are two basic types of moving iron meter, one based on magnetic attraction and the other on magnetic repulsion. With the attraction type (Figure 14.9) a pivoted soft iron disc is attracted by a coil when a current passes through the coil. The torque resulting from this attraction depends on both the current and the shape of the disc, the torque however being proportional to the square of the current. It is opposed by a torque resulting from springs, the torque being proportional to the angle through which the disc and pointer have rotated. Thus the angular deflection is proportional to the square of the current.

The repulsion type (Figure 14.10) has two pieces of soft iron inside a coil. One of the pieces is fixed and the other able to move, being fixed to the end of a pivoted pointer. When a current passes through the coil both pieces of iron become magnetized in the same way and so repulsion occurs, the amount of repulsion depending on the size of the current. The deflecting torque is proportional to the square of the current and is opposed by a restoring torque provided by springs, this being proportional to the angle through which they are twisted. The result is that the angular deflection is proportional to the square of the current. Damping is usually provided by an air damper involving a piston moving in a cylinder.

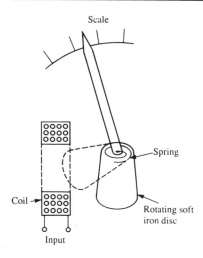

Figure 14.9 Attraction type moving iron meter

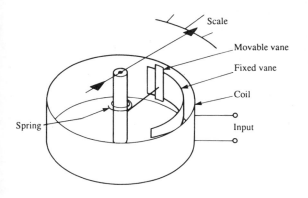

Figure 14.10 Repulsion type moving coil meter

14 Current

Since the deflection of the instrument is proportional to the square of the current the direction of the current has no effect on the reading. The instrument can thus be used for both d.c. and a.c. The scale is non-linear, being most cramped at the lower end, and for a.c. gives the r.m.s. current. Typically they have ranges with full scale deflections between 0.1 and 30 A, without shunts, and an accuracy of about ±0.5% of full scale deflection. When used with d.c. the instrument gives a low indication of a slowly increasing current and a high indication of a slowly decreasing current, this being because of the magnetic hysteresis curve of the iron used for the vanes. When used with a.c. resistive shunts can lead to errors as a result of changes in the coil reactance with frequency. This effect can be reduced by using a 'swamp' resistor in series with the instrument, or using a capacitor shunt, or a shunt with an inductance-to-resistance ratio equal to that of the coil. Errors can also be introduced if the waveform is non-sinusoidal. The instrument can be used with a.c. up to a frequency of about 100 Hz.

15 Voltage

As a voltmeter the moving iron meter has a relatively low input impedance, of the order of 50 Ω/V, and a minimum full scale deflection of about 50 V. It can be used for both direct and alternating voltages.

Thermocouple meter

The thermocouple meter involves a thermocouple being used to give a measure of the temperature of a resistor when a current passes through it. The output from the thermocouple can then be monitored by a d.c. moving coil meter. Since the heating effect of the current is proportional to the square of the current through the resistor or the square of the potential difference across it, the instrument can be used for alternating currents or voltages.

16 Current and voltage, a.c.

The instrument measures the true r.m.s. values, regardless of waveshape and can be used from about 10 Hz to 50 MHz. The scale is non-linear with full scale deflections between about 2 mA and 50 mA, and is used with a series resistor for voltages. The instrument is fragile with a low overload capacity.

Electrodynamometer

Figure 14.11 shows the principle of the electrodynamometer. It consists of two fixed coils to provide the magnetic field which causes the moving coil to rotate when a current passes through it. The magnetic flux density B, produced by a current I_1 through the two fixed coils, is proportional to I_1. The torque acting on the moving coil when it carries a current I_2 is proportional to BI_2. Hence the torque is proportional to I_1I_2. This deflecting torque results in the coil rotating against the restoring torque provided by springs. Since the restoring

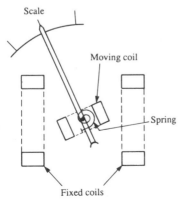

Figure 14.11 Dynamometer

torque is proportional to the angle θ through which the coil rotates then at equilibrium

$$\theta = kI_1I_2$$

where k is some constant. Stray magnetic fields can affect the operation of the instrument and therefore the coil system is enclosed in a magnetic shield.

17 Current

With the fixed and moving coils connected in series (Figure 14.12), i.e. $I_1 = I_2$, then the angle through which the moving coil rotates is proportional to the square of the current. The instrument can be used for the measurement of both d.c. and a.c., giving r.m.s. values irrespective of waveform. The arrangement shown in Figure 14.12 can be used for currents with full scale deflections between about 5 mA and 100 mA. For larger currents, up to about 20 A, the moving coil is shunted by a low resistance (Figure 14.13). Larger values of alternating current can be determined by using a current transformer with the ammeter. The instrument is more expensive than a moving coil instrument and has a higher power consumption.

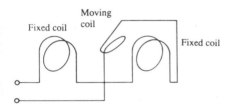

Figure 14.12 Dynamometer as a milliammeter

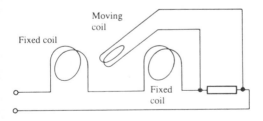

Figure 14.13 Dynamometer as an ammeter

18 Voltage

The form of the dynamometer referred to above for the measurement of current can be converted to a voltmeter by the addition of a series resistor. It can be used for both direct and alternating voltages up to about 2 kHz, giving r.m.s. values regardless of waveform. It has a non-linear scale.

19   Power

When used for the measurement of power dissipation the stationary coils are connected in series with the measurement circuit load, while the moving coil with a series resistor is connected in parallel with the load (Figure 14.14). Since the angular deflection of the movable coil is proportional to the product of the currents in the fixed and movable coils, and because the current in the stationary coil is the current through the load and the current through the moving coil is proportional to the potential difference across the load, the angular deflection is proportional to the product of the current and potential difference. Theoretically the instrument thus responds to the instantaneous power but system inertia means that the instrument responds to the average power. Such an arrangement is known as a *wattmeter*.

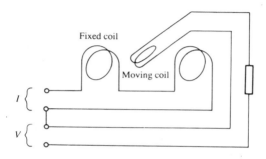

Figure 14.14   Dynamometer as a wattmeter

Two ways of connecting the wattmeter into a circuit are shown in Figure 14.15. In Figure 14.15(a) the moving coil gives a measure of the potential difference across the fixed coils and the load with the fixed coils measuring the current through just the load while in Figure 14.15(b) the potential difference is just that across the load and the current is that through the movable coil and the load. With the (a) circuit the wattmeter reads high by the power due to the potential drop across the fixed coils, i.e. $I_L^2 R_f$ with I_L being the load current and R_f the resistance of the fixed coils. With the (b) circuit the wattmeter reads high by the power due to the current through the movable coil, i.e. V_L^2/R_m with V_L being the potential difference across the load and R_m the resistance of the movable coil. Generally circuit (a) is preferred for low current–high voltage loads and (b) for high current–low voltage loads.

A compensated wattmeter which avoids the necessity for making the corrections outlined above is shown in Figure 14.16. The fixed coils each have two windings, each with the same number of turns. One winding uses heavy gauge wire and carries the load current, the other with finer gauge wire carries only the current to the movable coil, i.e. the current through the voltage coil which is small because of the resistance in series with that coil. This voltage coil current is however in the opposite direction to the load current through the fixed coils and cancels out the proportion of the magnetic flux due to the voltage coil current. The wattmeter thus indicates the correct power.

Wattmeters can be used for products of voltages up to about 300 V and currents of 20 A, for d.c. and a.c. up to frequencies of 400 Hz, with an

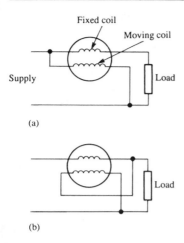

(a)

(b)

Figure 14.15 Wattmeter connections

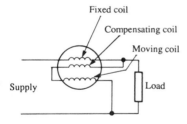

Figure 14.16 Compensated wattmeter

accuracy of about ± 0.1 to 0.5% of full scale deflection. The meter is not suitable for measuring powers below about 5 W.

20 Power factor

The power factor is the cosine of the angle between the voltage and the current. The *crossed coil power factor meter* consists of a dynamometer with the moving element as two coils, mounted on the same shaft but at right angles to each other (Figure 14.17). One of the coils has an inductor in series with it and is connected across the load, the other has a resistor in series with it and it also is connected across the load. The currents in the two coils are equal in magnitude but displaced in time by 90°. No restoring torque springs are used, the angular position of the moving element being determined by the resultant torque developed by the two crossed coils. The resulting angular displacement is a measure of the phase angle between the voltage and the current.

Electronic wattmeter

21 Power

The electronic wattmeter essentially multiplies the inputs corresponding to the load current and potential difference and then takes an

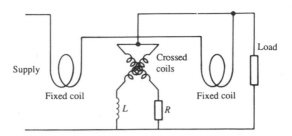

Figure 14.17 Crossed coil power factor meter

average value over a period of time and displays it. This process may be continuous or employ sampling. There are a number of methods used to give multiplication with a continuous display of the power. One method uses pulses, the current being used to determine the width of the pulse and the voltage the height. The average pulse area over a period of time thus becomes a measure of the average power. The sampling method involves samples being simultaneously taken of the voltage and current. These are then converted to digital form and multiplied and averaged by digital circuits. Typically the electronic wattmeter measures powers over the range 0.1 W to 100 kW at frequencies up to 100 kHz. Its accuracy tends to decrease with frequency, from about $\pm 0.5\%$ at low frequencies to $\pm 1\%$ at high frequencies. A continuous electronic wattmeter employing the Hall effect can be used at frequencies up to several GHz.

Watthour meter

22 Energy

The watthour meter (Figure 14.18) is widely used for the measurement of the electrical energy supplied to industrial and domestic consumers through the mains alternating supply. A light aluminium disc which is free to rotate in a horizontal plane passes at its rim through the air gap of a magnetic circuit, the flux in the circuit being produced by the current through the load. The disc also passes through the air gap in a second magnetic circuit, the flux in this circuit being produced by the potential difference across the load. The flux generated by the voltage is arranged, by means of the copper ring in the voltage magnetic circuit, to be exactly $90°$ out of phase with that in the current magnetic circuit. Since the current and the voltage are alternating the flux in each of the magnetic circuits is alternating and so eddy currents are induced in the aluminium disc. The interaction between these currents and the magnetic field in which they are located causes the disc to rotate. The average torque acting on the disc is proportional to $VI \cos \phi$. This torque is opposed by a permanent magnet acting as a brake by inducing eddy currents in the moving disc, and the interaction between these and the magnet producing a braking effect which is proportional to the speed of rotating of the disc. At equilibrium the generated and braking torques are equal and so the speed of rotation of the disc is proportional to $VI \cos \phi$. The shaft of the disc is connected via gearing to a mechanical counter which thus gives a count proportional to the watt-hours.

Electrostatic meters

Electrostatic instruments depend for their action on the forces that occur between two charged bodies.

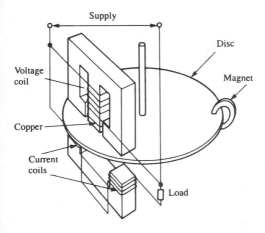

Figure 14.18 Watthour meter

23 *Voltage*

Figure 14.19 shows the basic form of the electrostatic voltmeter. It consists of a set of four quadrant-shaped boxes in which a movable vane can rotate. Figure 14.19 shows the most commonly used form of connection, called *idiostatic*. Opposite boxes are connected together, with the movable vane connected to one pair, so that they become oppositely charged by the voltage to which they are connected. As a

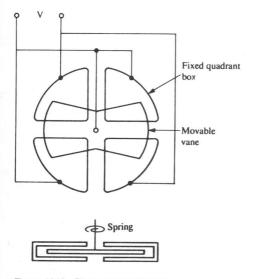

Figure 14.19 Electrostatic voltmeter

consequence the movable vane experiences a torque which is proportional to the square of the voltage. A restoring torque proportional to the angular movement of the vane is provided by springs. Thus at equilibrium the angular deflection of the vane is proportional to the square of the voltage. Because of this relationship the instrument can be used for direct or alternating voltages, usually being scaled for r.m.s. values. Such instruments have a high input impedance, are fragile, accurate and expensive. Multicell arrangements are possible. The range of such instruments tends to be from about 100 V to 1000 V or higher.

Digital voltmeters

24 Voltage d.c.

Digital voltmeters (see Chapter 10) can be classified according to the form of analogue to digital converter (see Chapter 9, item 25) they use, the most common being *successive approximations*, *ramp*, *dual ramp* and *pulse width*. The successive approximations and ramp forms are examples of sampling meters in that they provide digital values equivalent to the voltage at a particular time instant. The dual ramp and pulse width forms are examples of integrating meters in that the average value of the voltage is given over a fixed measurement time. Integrating forms take longer to carry out a measurement but have better noise rejection.

With the *successive approximations* form a sample of the input voltage is compared with a voltage which is increased in increments until its total value equals the input voltage. Typically an 8-bit meter will have a conversion time of about $10 \mu s$, a maximum of 16 bits being used for this form. Sampling times are typically of the order of 1000 times or more per second.

The successive approximations form of digital voltmeter is one of the faster responding voltmeters. For very fast responses, instead of comparing the input voltage with steadily mounting voltage increments and building up to the required voltage the comparison can be made simultaneously with a large range of voltages, each one being linked to a digital code, and the matching voltage rapidly found. Such a form of digital voltmeter is said to employ a *flash converter* and has conversion times of the order of 10 ns.

The *ramp* form is the simplest and cheapest form of digital voltmeter, the input voltage being compared with a steadily increasing ramp voltage. The time between the two voltages being equal and the end of the ramp voltage is a measure of the input voltage. Because of non-linearities in the shape of the ramp waveform used and its lack of noise rejection, accuracy is limited to about $\pm 0.05\%$. Sample rates can be up to about 1000 times per second.

The *dual ramp* form involves a capacitor being charged during a time equal to 1 cycle of the line frequency. The resulting potential difference is then compared with a steadily increasing ramp voltage and the time taken for the two to become equal is a measure of the input voltage. It has the advantage of noise and line-frequency signal rejection but since it integrates the signal over 1 cycle of the mains frequency, has a conversion time of only the reciprocal of the mains frequency, i.e. 1/50 or 1/60 s. Accuracy is about $\pm 0.005\%$.

Pulse width forms produce pulses whose width, i.e. duration, is proportional to the input voltage. The duration of the pulse is then measured by a clock. By integrating over 1/50 or 1/60 s rejection of the line frequency occurs and high resolution is possible.

Digital meters provide a numerical readout which eliminates interpolation and parallax errors. The resolution of such an instrument corresponds to the voltage change which gives a change in the least significant bit of the meter display. Displays are generally between $3\frac{1}{2}$ and $8\frac{1}{2}$ digits, the half being because the most significant digit can only take the value 0 or 1. A $3\frac{1}{2}$ digit display has a resolution of 1 in 2000 and a $8\frac{1}{2}$ digit display 1 in 2×10^8. Typically a $3\frac{1}{2}$ digit meter will have an accuracy of $\pm 0.1\%$ of the reading plus 1 digit, while a $8\frac{1}{2}$ digit display accuracy is 0.0001% of the reading plus 0.00003% of the full scale reading. Typically such instruments have input resistances of 10 MΩ or higher, capacitances of 40 pF, and good stability. Voltage ranges vary from about 100 mV to 1000 V with the limit of resolution being about 1 μV.

25 Voltage a.c.
The methods used to convert a.c. to d.c. are similar to those used with permanent magnet moving coil instruments (see earlier this chapter). Thus rectification methods give average values and since the instruments are generally scaled to read r.m.s. values the result needs correction for non-sinusoidal waveforms. True r.m.s. readings can be obtained using a thermocouple to monitor the temperature of a resistor across which the input voltage is applied or using electronics to square the voltage and then extract the square root.

Typically accuracy varies from about $\pm 1\%$ of the reading plus 3 digits with a $3\frac{1}{2}$ digit display to $\pm 0.05\%$ of the reading plus 0.03% of the full scale reading for a $8\frac{1}{2}$ digit display. The frequency range varies from about 45 Hz to 10 kHz for a $3\frac{1}{2}$ digit display to 10 Hz to 100 kHz for a $8\frac{1}{2}$ digit display. The input impedance is typically about 10 MΩ with 100 pF. Voltage ranges vary from full scale readings of about 100 mV to 1000 V r.m.s.

26 Current
Both d.c. and a.c. currents are determined by the digital voltmeter being used to measure the potential difference across a standard resistor. Typically the accuracy is about $\pm 0.2\%$ of the reading plus 2 digits for d.c. and $\pm 1\%$ of the reading plus 2 or more digits for a.c. For both d.c. and a.c. the ranges are from about 200 μA to 2 A and the voltage drop less than 0.3 V. The frequency range is about 45 Hz to 1 kHz.

27 Resistance
Resistances can be measured using a digital voltmeter by passing a known current through the resistance and then determining the resulting potential difference across it. Higher accuracy is, however, obtained by passing the same current through a standard resistor and the unknown resistor and comparing the potential differences across the two. Accuracy varies from about $\pm 0.1\%$ of the reading plus 1 digit for a $3\frac{1}{2}$ digit meter to $\pm 0.0002\%$ of the reading plus $\pm 0.0004\%$ of the full scale reading for a $8\frac{1}{2}$ digit display. The resistances ranges are from about 200 Ω to 1000 MΩ.

DC bridges

28 Wheatstone bridge
For details of the Wheatstone bridge see Chapter 9 and item 1. Such bridges, in the form shown in Figure 9.1, are typically used for the measurement of resistance in the range 1 Ω to 10 MΩ.

29 Kelvin double bridge
Low resistances can be measured using the Kelvin double bridge, as in Figure 14.20. R_1 is the resistance being measured and R_2 a standard

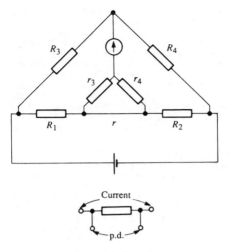

Four terminal resistor

Figure 14.20 Kelvin double bridge

resistance of about the same size. Because such resistors have low resistance it is necessary to accurately define the resistance by using four-terminal resistors. Two of the terminals define the points between which the current is supplied and two the points between which the potential difference is determined, so avoiding uncertain contact resistance problems. The resistance of the connector linking the two resistors, sometimes referred to as the yoke, is r. R_3, R_4, r_3 and r_4 are resistances for which either R_3 and r_3 or R_4 and r_4 are variable and the relationship between their resistances is given by

$$\frac{R_3}{R_4} = \frac{r_3}{r_4}$$

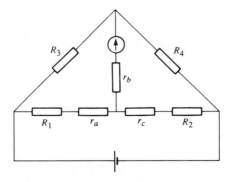

Figure 14.21 Equivalent circuit

If the delta star transformation is applied to the double bridge the equivalent circuit shown in Figure 14.21 is obtained with

$$r_a = \frac{r_3 r}{r_3 + r_4 + r}$$

$$r_c = \frac{r_4 r}{r_3 + r_4 + r}$$

The balance condition is thus

$$\frac{R_1 + r_a}{R_2 + r_c} = \frac{R_3}{R_4}$$

Hence

$$R_1 = \frac{R_3(R_2 + r_c)}{R_4} - r_a$$

$$= \frac{R_3 R_2}{R_4} + \frac{R_3 r_4 r}{r_3 + r_4 + r} - \frac{r_3 r}{r_3 + r_4 + r}$$

$$= \frac{R_3 R_2}{R_4} + \frac{r_4 r}{r_3 + r_4 + r} \left(\frac{R_3}{R_4} - \frac{r_3}{r_4} \right)$$

But R_3/R_4 has been made equal to r_3/r_4, hence the balance condition is

$$R_1 = \frac{R_3 R_2}{R_4}$$

The bridge is used for measuring resistances in the range $0.1\ \mu\Omega$ to $1\ \Omega$.

30 High resistance bridge

Figure 14.22 shows a modified form of Wheatstone bridge that can be used for the measurement of high resistances. The measurement of high resistance presents problems because of parallel leakage paths and thus three-terminal resistances are used. Such resistances have the resistor mounted on insulating pillars above a metal plate, two terminals being directly connected to the resistor and the third to the plate. There will be leakage resistances between each resistor terminal and the plate. When the resistor R_1 is connected into the bridge the metal plate is connected to the detector junction with R_3 and R_4. This puts the leakage resistance R_{s1} in parallel with R_3 and since R_{s1} is much larger than R_3 its effect is negligible. The leakage resistance R_{s2} is in parallel with the detector and its only effect is to change the detector sensitivity.

AC bridges

There are many forms of a.c. bridges, the following being just some of the more commonly used ones for the measurement of capacitance and inductance. See Chapter 9 for basic bridge theory and circuits for a number of bridges. The detectors used to determine the balance condition include earphones (from about 250 Hz to 3 kHz), a vibration galvanometer (a light spot galvanometer which is tuned to have a natural frequency at the supply frequency), and a tunable amplifier (tuned to respond to a narrow bandwidth at the frequency used by the bridge which can be between about 10 Hz and 100 kHz).

Further reading: Hague, B. and Foord, T. R. (1971), *Alternating Current Bridges*, Pitman.

31 Inductance

The following are examples of bridges that are used for the measurement of the inductance and resistance of an inductor. See Chapter 9 for

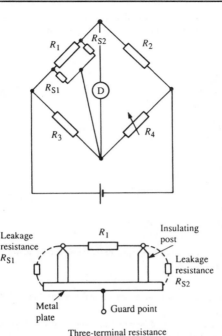

Figure 14.22 High resistance bridge

details. The *Maxwell* bridge (Figure 9.8(b)) is used with inductors having high Q-factors. The *Maxwell–Wien* bridge (Figure 9.7) is limited to the measurement of inductors with low Q-factors. The *Owen* bridge (Figure 9.8(a)) is used with inductors having a large inductance. The *Hay* bridge (Figure 9.8(c)) is used with very high Q-factors.

32 Capacitance

The following are examples of bridges used for the measurement of capacitance. See Chapter 9 for details. The *De Sauty* bridge (Figure 9.9(a)) measures a capacitance by comparison with another known capacitance, however the power factor of the capacitor is significant and so the bridge is usually inadequate in many cases. The *Schering* bridge (Figure 9.9(c)) is used for the measurement of the parallel capacitance and resistance of a capacitor and hence the dielectric losses. The *Wien* bridge (Figure 9.9(b)) has a number of applications. It can be used for the measurement of capacitance if the frequency of the supply is known, or conversely the frequency if the capacitance is known. It is also used in a number of applications as a means of discriminating against a particular frequency.

Potentiometer bridges

The potentiometer is a null method based on opposing the unknown potential difference with another so that they cancel and no current flows.

33 EMF
The basic potentiometer circuit described in Chapter 9 and shown in Figure 9.12 can be used for the comparison of an unknown e.m.f. or potential difference with a known e.m.f. The standard cell often used is the Weston cell and this has an e.m.f. of 1.01880 V at 15 °C, 1.01862 at 20 °C. For very small e.m.f.s, e.g. those from a thermocouple, the adaptation shown in Figure 9.13 can be used.

Transformer bridges

34 Single ratio transformer bridge
Figure 14.23 shows the basic form of a single ratio transformer bridge, a tapped transformer providing a voltage division which is dependent on the number of turns across which the tapping is made. The voltage across N_1 turns is proportional to N_1, and that across N_2 proportional to N_2. Thus

$$V_1 = kN_1 = I_1 Z_1$$
$$V_2 = kN_2 = I_2 Z_2$$

where k is a constant. Thus $I_1 = kN_1/Z_1$ and $I_2 = kN_2/Z_2$. At balance there is zero current through the detector. This means I_1 must equal I_2, hence

$$\frac{N_1}{Z_1} = \frac{N_2}{Z_2}$$

The bridge can be used for the measurement of the resistance, capacitance and inductance of components over a wide range of frequencies. For resistance measurements Z_1 can be the unknown resistance R_x and Z_2 a standard resistance R_s. Thus the ratio of the resistances when the current through the detector is zero is the ratio of the turns tapped.

$$R_x = \frac{N_1 R_s}{N_2}$$

For capacitance measurement Z_1 can be the unknown capacitor, capacitance C_x with effectively parallel resistance of R_x, and Z_2 a standard capacitor C_s with a parallel variable resistance R_s. At balance

$$R_x = \frac{N_1 R_s}{N_2}$$

$$C_x = \frac{N_2 C_s}{N_1}$$

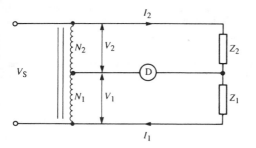

Figure 14.23 Single ratio transformer bridge

35 Double ratio transformer bridge

Figure 14.24 shows the basic form of the double ratio transformer bridge. When the detector indicates zero current then there is zero magnetic flux in the core of the second transformer. For this to occur the ampere turns $n_1 I_1$ and $n_2 I_2$ must be equal and producing fluxes which oppose each other. But $I_1 = V_1/Z_1$ with $V_1 = kN_1$, and $I_2 = V_2/Z_2$ with $V_2 = kN_2$. Hence

$$\frac{n_1 N_1}{Z_1} = \frac{n_2 N_2}{Z_2}$$

The use of the two turns ratios enables a small number of standard components to be used for a greater range of measurements than is the case with the single ratio bridge.

The input transformer used is generally a multi-decade ratio transformer. Such a transformer has sets of tappings, each of which has nine tappings with the first decade being for n turns per tap, the second decade for $10n$ turns per tap, the third for $100n$ turns per tap, etc. Thus by using these different tappings with a few standard components a finely and widely variable scale for measurement is produced. The bridge can be used for the measurement of resistance, capacitance and inductance at frequencies up to about 250 MHz.

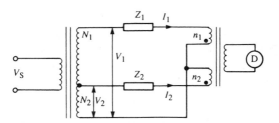

Figure 14.24   Double ratio transformer bridge

Q-meter

The *Q-factor* is defined as

$$Q = \frac{2\pi \times \text{maximum stored energy per cycle}}{\text{energy dissipated per cycle}}$$

For a series resonant circuit (Figure 14.25) with an inductor L and capacitor C the Q-factor is

$$Q = \frac{\omega_0 L}{R} = \frac{1}{\omega_0 C R}$$

where ω_0 is the resonant frequency. At this resonant frequency the voltage V_c across the capacitor is

$$V_c = i_{res} \times \frac{1}{\omega_0 C}$$

The resonant current i_{res} is however only determined by the circuit resistance R, being therefore V_s/R. Hence

$$V_c = V_s \times \frac{1}{\omega_0 C R} = V_s \times \frac{\omega_0 L}{R} = V_s Q$$

The voltage across the capacitor is thus the Q-factor multiplied by V_s.

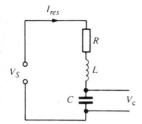

Figure 14.25 Series resonant circuit

36 Impedance

Figure 14.26 shows the basic Q-meter circuit, it essentially being just the series resonant circuit. An oscillator passes current through a very low resistance element, of the order of $0.02\,\Omega$, in the resonant circuit. This acts as a voltage source V_s with a very small internal resistance. This voltage is generally measured by a thermocouple meter which has a scale giving the factor by which the V_C value must be multiplied. The voltage across the variable capacitor V_C is measured by an electronic voltmeter which has a scale directly giving Q-factor values. The figure shows the arrangement for the measurement of the Q-factor for an unknown inductor and hence the inductance, though corrections may have to be made for the self capacitance of the inductor.

Low impedances, e.g. large capacitances, small inductances and low resistances are determined by connecting the unknown component such as a capacitor C_x in series with the variable capacitor and inductor (Figure 14.27(a)). Firstly the unknown capacitor is short circuited and the circuit tuned to give a Q-factor value. If the variable capacitor for this has a value C_1 and the frequency is ω_0 then

$$Q_1 = \frac{1}{\omega_0 C_1 R}$$

Then the short circuit is removed and the circuit tuned at the same frequency as before by adjusting the variable capacitor to a value C_2 to give a Q-factor of Q_2. The capacitance in the circuit is now C_2 in series with C_x. Since the same resonant frequency is used this capacitance must be equal to C_1. Hence

$$\frac{1}{C_1} = \frac{1}{C_x} + \frac{1}{C_2}$$

$$C_x = \frac{C_1 C_2}{C_2 - C_1}$$

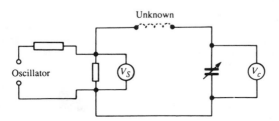

Figure 14.26 Q-meter

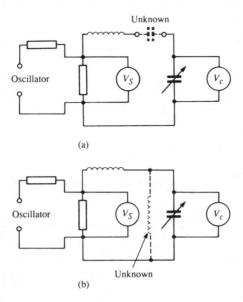

(a)

(b)

Figure 14.27 Q-meter (a) series arrangement (b) parallel arrangement

The total resistance in the circuit is R plus the leakage resistance R_x which can be assumed to be in parallel with the unknown capacitor. Hence

$$R_x = \text{total resistance} - R$$

But $R = 1/\omega_0 C_1 Q_1$ and the total resistance $= 1/\omega_0 C_2 Q_2$. Hence

$$R_x = \frac{C_1 Q_1 - C_2 Q_2}{\omega_0 C_1 C_2 Q_1 Q_2}$$

The Q-factor for the unknown capacitor Q_x is $1/\omega_0 R_x C_x$ and so

$$Q_x = \frac{Q_1 Q_2 (C_1 - C_2)}{C_1 Q_1 - C_2 Q_2}$$

If the unknown component had been a small inductor instead of large capacitor then

$$L_x = \frac{C_1 - C_2}{\omega_0^2 C_1 C_2}$$

If the unknown component had been a pure resistor then C_1 would equal C_2 since there would have been no change in resonance condition, hence

$$R_x = \frac{Q_1 - Q_2}{\omega_0 C_1 Q_1 Q_2}$$

For high impedance components, e.g. small capacitors, some inductors and large resistors, the unknown component is connected in parallel with the variable capacitor (Figure 14.27(b)). Firstly with the component absent the resonance condition is determined with the variable capacitor as C_1 and the Q-factor as Q_1. Then with the component connected the variable capacitor is adjusted to give resonance at the same frequency. If the variable capacitor is then C_2 and the Q-factor Q_2 then if the unknown is capacitive C_x

$$C_x = C_1 - C_2$$

If the unknown is inductive L_x,

$$L_x = \frac{1}{\omega_0^2(C_1 - C_2)}$$

The resistance R_x of the unknown impedance is

$$R_x = \frac{Q_1 Q_2}{\omega_0 C_1(Q_1 - Q_2)}$$

The Q-factor of the unknown component Q_x is

$$Q_x = \frac{Q_1 Q_2(C_1 - C_2)}{C_1(Q_1 - Q_2)}$$

Frequency measurement

37 Digital frequency counter

Figure 14.28 shows the basis of a digital frequency counter. A quartz crystal oscillator is used to provide a stable frequency. The oscillator may be uncompensated, temperature compensated or oven stabilized, the extent of the compensation determining the accuracy possible and hence the number of digits the instrument has. In the conventional form of frequency counter a gate is opened for a time determined by n cycles of the oscillator frequency f, i.e. a time n/f. During that time the

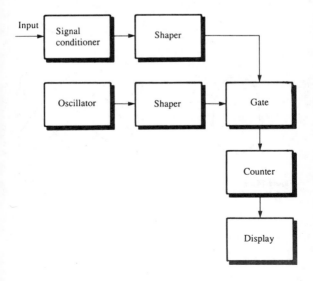

Figure 14.28 Digital frequency counter

number of cycles n_i of the input signal are counted. The frequency of the input signal is thus $n_i/(N/F)$. A typical uncompensated oscillator will give an instrument with five or six digits an an accuracy of 1 part in 10^5 or 10^6. An oven stabilized oscillator will give seven to nine digits and an accuracy of 1 part in 10^7 to 10^9. Such instruments have typically frequency ranges from d.c. to 200 MHz.

38 Lissajous figures

This technique is used to compare the frequency or phase of one signal with that of another. One signal is fed to the Y plates of an oscilloscope and the other to the X plates. The pattern obtained on the screen depends on the ratio of the two frequencies and their phase difference. The frequency ratio is determined by counting the number of peaks in the X-direction n_x and the number of peaks in the Y-direction n_y, (Figure 14.29).

$$F_y n_y = f_x n_x$$

If two signals are of the same frequency then the pattern obtained can be of the form shown in Figure 14.30 with the phase difference θ between the two signals being given by

$$\sin \theta = \frac{y_0}{y_{max}}$$

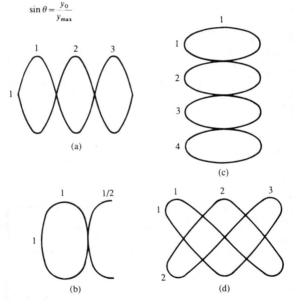

Figure 14.29 Lissajous figures: (a) $f_y/f_x = 3/1$, (b) $f_y/f_x = 1\frac{1}{2}/1 = 3/2$, (c) $f_y/f_x = 1/4$, (d) $f_y/f_x = 3/2$

Figure 14.30 Phase difference

15 Flow

The term *fluid* is used to describe substances that flow, i.e. both liquids and gases. When a fluid flows with *orderly motion*, sometimes referred to as streamline motion or laminar flow, through a tube or past a surface, every particle of the fluid moves in straight lines parallel to the tube walls. However the fluid immediately adjacent to the walls is slowed down as a result of viscosity, at the tube walls dropping down to zero. A consequence of this is that there is a velocity gradient (Figure 15.1(a)). This represents the situation below a critical rate of flow, at higher rates of flow the motion becomes chaotic, each fluid particle following a very irregular path. Such motion is said to be *turbulent flow*. Despite the chaotic nature of the flow the average result for flow along a tube is like that shown in Figure 15.1(b). Orderly flow through pipes can be expected if the *Reynolds number* is less than 2000 and turbulent if more than 4000. Between these values the transition occurs.

$$\text{Reynolds number} + \frac{Dv\rho}{\eta}$$

where D is the pipe diameter, v the fluid velocity, ρ the fluid density and η its viscosity.

If the average velocity of a fluid through a pipe is v then in a time t the flow will have advanced a distance vt. If the cross-sectional area of the tube is A then the volume of fluid that has moved through this distance in time t is Avt. The volume rate of flow Q is thus (Avt/t). Thus

$$Q = Av$$

A fluid in motion has potential energy, kinetic energy, pressure energy and heat energy, its total energy being the sum of all these terms.

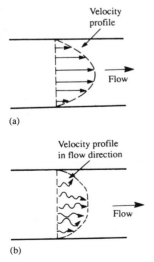

(a)

(b)

Figure 15.1 (a) Orderly flow (b) turbulent flow

Flow measurement can be considered to include measurements of three different quantities, the actual velocity of the fluid at some point in the fluid, the volume rate of flow, and the mass rate of flow. Table 15.1 lists the flow measurement systems discussed in this chapter and their characteristics.

Further reading: Noltingk, B. E. (ed.) (1985), *Jones' Instrument Technology*, vol. 1 (*Mechanical Measurements*), Butterworth-Heinemann.

Table 15.1 Flow measurement systems

Principle	System	Characteristics
Volume flowrate		
Differential pressure	1 Venturi tube	Simple, accurate, reliable, pressure loss 10 to 15%, non-linear, can be used for dilute slurries.
	2 Orifice plate	Simple, cheap, accuracy $\pm 1.5\%$ pressure loss 50 to 70%, non-linear.
	3 Nozzles	Cheaper than venturi but more than orifice plate, accuracy $\pm 0.5\%$, pressure loss 40 to 60%.
	4 Dall tube	Larger pressure difference and more compact than venturi, pressure loss 4 to 6%.
	5 Variable area	Rotameter: accuracy $\pm 1\%$, cheap, range 30×10^{-6} to $1\,m^3/s$.
	6 Variable orifice	Linear, water range up to $3\,m^3/s$, up to 200 bar pressure and $500\,°C$.
Mechanical	7 Target	Range up to about $0.03\,m^3/s$, accuracy $\pm 0.5\%$, can be used for viscous and dirty fluids.
	8 Turbine	Range up to about $1\,m^3/s$, accuracy $\pm 0.3\%$, good repeatability, expensive.
	9 Rotating vane	Used for liquids and gases.
Positive displacement	10 Rotary piston	Accuracy $\pm 1\%$, used for water.
	11 Reciprocating piston	Accuracy $\pm 0.1\%$, wide range, used for liquids.
	12 Nutating disc	Accuracy $\pm 1\%$, used for liquids.
	13 Rotating impeller	Fluted: use for oil up to $1\,m^3/s$ and 80 bar pressure. Lobed: used for gases $0.003\,m^3/s$ to $3\,m^3/s$, accuracy $\pm 1\%$.
	14 Rotating vane	Used for oil and fuel, accuracy $\pm 0.1\%$.

continued

Table 15.1 (*continued*)

Principle	System	Characteristics
	15 Diaphragm meter	Used for metering domestic gas.
	16 Liquid sealed drum	Used for gas flow.
Electromagnetic	17 Electromagnetic	Used for conductive liquids, speeds up to 10 m/s, accuracy ±1%.
Ultrasonic	18 Doppler	Relatively cheap, poor accuracy (±5% or worse), useful as flow indicator or switch.
	19 Time of flight	Used for fluids in pipes and open channels, speeds from 0.2 m/s to 12 m/s, accuracy ±1%.
	20 Cross-correlation	Slow response time.
Oscillatory	21 Vortex	Used for both liquids and gases, at high pressures and temperatures, accuracy ±1%.
	22 Swirl meter	Used for liquids, range 6×10^{-4} m³/s to 2 m³/s, and gases, 10^{-3} m³/s to 3 m³/s, accuracy ±1%.
Mass flowrate Direct measurement	23 Coriolis	Used for liquids or gases, accuracy ±0.5%.
	24 Thermal	Used for gases in the range 2.5×10^{-10} kg/s to 5×10^{-3} kg/s, accuracy ±1%.
Inferential	25 Turbine–vibrating element	Used for liquids and gases.
Point velocity Pressure	26 Pitot tube	For liquids or gases, accuracy ±1 or 2%.
	27 Annubar	For liquids or gases, accuracy ±1%.
Thermal	28 Hot wire anemometer	For gases range 0.1 m/s to 500 m/s, liquids 0.01 m/s to 5 m/s, accuracy ±1%.

Differential pressure flowmeters

When a fluid flows through a constriction in a pipe it accelerates and its velocity increases, i.e. there is a gain in kinetic energy. For a horizontal pipe this energy results in a drop in pressure energy and so the pressure drops. For flow which is incompressible, i.e. the density does not change when the pressure changes, applying the conservation of energy with a horizontal pipe where v_1 is the fluid velocity, P_1 the pressure and A_1 the cross-sectional area of the pipe, v_2 the velocity, P_2 the pressure and A_2 the cross-sectional area at the constriction, and ρ the fluid density:

$$\frac{v_1{}^2}{2g} + \frac{P_1}{\rho g} = \frac{v_2{}^2}{2g} + \frac{P_2}{\rho g}$$

The equation is known as *Bernoulli's equation*. Since the density does not change the volume of fluid Q passing through the wide section per second must equal the volume passing through the constriction. Hence

$$Q = A_1 v_1 = A_2 v_2$$

where A_1 is the cross-sectional area of the tube and A_2 that at the constriction. Hence

$$Q = \frac{A_2}{\sqrt{[1-(A_2/A_1)^2]}} \sqrt{[2(P_1 - P_2)/\rho]}$$

In practice this equation is only an approximation and is modified by a correction factor C, called the *discharge coefficient*, to

$$Q = \frac{CA_2}{\sqrt{[1-(A_2/A_1)^2]}} \sqrt{[2(P_1 - P_2)/\rho]}$$

C is a function of pipe size, the Reynolds number for the flow, and the form of instrument used to measure the pressure difference. Tables and equations are available (British Standards Institution, BS 1042: section 1.1, 1981 and International organization for Standardization, ISO 5167, 1980) to enable C values to be determined for particular configurations.

The mass flow rate for an incompressible fluid is $q\rho$ and thus

$$\text{mass flow rate} = \frac{CA_2}{\sqrt{[1-(A_2/A_1)^2]}} \sqrt{[2\rho(P_1 - P_2)]}$$

The term E, called the *velocity of approach factor*, is often used in both the above equations for

$$E = \frac{1}{\sqrt{[1-(A_2/A_1)^2]}}$$

The above refers to incompressible fluids. For compressible fluids such as gases, when the pressure changes there is a density change. For an adiabatic change PV^γ is a constant, with γ being the ratio of the specific heats at constant pressure and constant volume. Thus

$$\frac{P_1}{\rho_1{}^\gamma} = \frac{P_2}{\rho_2{}^\gamma}$$

where ρ_1 and ρ_2 are the densities and the pressures are respectively P_1 and P_2. Hence the conservation of energy equation becomes modified to

$$\frac{v_1{}^2}{2g} + \frac{\gamma}{\gamma-1}\frac{P_1}{\rho\gamma} = \frac{v_2{}^2}{2\gamma} + \frac{\gamma}{\gamma-1}\frac{P_2}{\rho g}$$

Mass is conserved in flowing from one diameter pipe to another and so

$$\text{mass flow rate} = v_1 \rho_1 A_1 = v_2 \rho_2 A_2$$

and thus

$$\text{mass flow rate} = \frac{C\varepsilon A_2}{\sqrt{[1-(A_2/A_1)^2]}} \sqrt{[2(P_1 - P_2)/\rho]}$$

where ε, called the *expansibility factor*, is given by

$$\varepsilon = \left[\left(\frac{\gamma r^{2/\gamma}}{\gamma - 1} \right) \left(\frac{1 - r^{(\gamma - 1)/\gamma}}{1 - r} \right) \left(\frac{1 - m^2}{1 - m^2 r^{2/\gamma}} \right) \right]^{\frac{1}{2}}$$

with $r = P_1/P_2$ and $m = A_2/A_1$. See the B.S. and I.S.O. references given above for more details. For incompressible fluids ε has the value 1.

There are a number of forms of flowmeter based on this principle of measuring the pressure difference resulting when a fluid flows through a constriction. All however have the problem of the non-linear relationship between quantity rate of flow and the pressure difference that is measured.

1 Venturi tube

The Venturi tube (Figure 15.2) has a gradual tapering of the pipe from the full diameter to the constricted diameter. The constricted diameter should be not less than $0.224D$ and not more than $0.742D$, where D is the full diameter. The inlet taper of the tube should be $10.5° \pm 1°$ and the exit taper between $5°$ and $15°$. The pressure difference between the flow prior to the constriction and at the constriction can be measured with a simple U-tube manometer or a diaphragm pressure cell. The discharge coefficient C is about 0.99 and the pressure loss occurring as a result of the presence of the Venturi tube is about 10 to 15%, a low value. It can be used with liquids containing particles, dilute slurries. It is simple in operation, is capable of high accuracy (can be better than $\pm 0.5\%$), has long term reliability, but is expensive and has a non-linear pressure–volume rate of flow relationship.

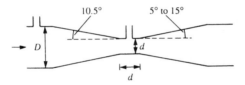

Figure 15.2 Venturi tube

2 Orifice plate

The orifice plate (Figure 15.3) is simply a disc with a hole, there being a number of forms. The most widely used is the concentric form with a central circular hole, with other forms being the eccentric with an off-centre circular hole, for use where condensed liquids are present in the gas flow or undissolved gases in a liquid flow, and the segmental with just a segment of the central circular hole, for use where particles are present in a liquid flow. The pressure difference can be measured between a point equal to the diameter of the tube upstream and a point a distance equal to half the diameter downstream, or at points on either side of the plate. It has a discharge coefficient C of about 0.6 and has a non-linear volume rate of flow pressure relationship. The orifice plate is simple, reliable, produces a greater pressure difference (more than twice as much) and is cheaper but less accurate ($\pm 1.5\%$ or more) and produces a greater pressure drop (about 50 to 70%) than the venturi tube. Problems of silting and clogging of the orifice can occur if particles are present in liquids.

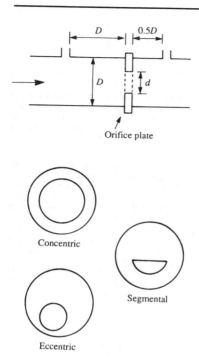

Concentric

Segmental

Eccentric

Figure 15.3 Orifice plate

3 Nozzle flowmeter

Two types of nozzle are used, the venturi nozzle and the flow nozzle. The venturi nozzle (Figure 15.4(a)) is effectively a venturi tube with the inlet cone considerably shortened. The flow nozzle (Figure 15.4(b)) is even shorter. Nozzles have a discharge coefficient C value of about 0.96 and produce pressure losses of the order of 40 to 60%. Nozzles are cheaper than venturi tubes, giving similar pressure differences, but more expensive than orifice plates and give a smaller pressure drop than orifice plates. They have an accuracy of about $\pm 0.5\%$, and have a non-linear volume-rate-of-flow/pressure relationship.

4 Dall flowmeter

This is a version of the venturi tube which gives a higher differential pressure (more than twice as much) and a lower pressure drop (about 4 to 6%). The Dall tube (Figure 15.5) is only about two pipe-diameters long. An even shorter form, a Dall orifice, is available. The Dall tube has a discharge coefficient C of about 0.66 and is often used where space does not permit the use of a venturi tube.

5 Variable area flowmeter

Variable area flowmeters involve maintaining a constant pressure difference by changing the area of the constriction through which the fluid flows. One form is essentially just an orifice plate in which the size

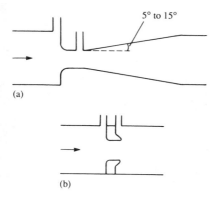

Figure 15.4 (a) Venturi nozzle (b) flow nozzle

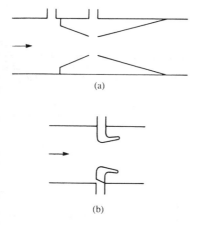

Figure 15.5 (a) Dall tube (b) Dall orifice

of the orifice is adjusted to give a constant pressure difference and is called a *variable area gate meter*. The most common form of variable area flowmeter is the *rotameter* (Figure 15.6). This has a float in a tapered vertical tube with the fluid flow pushing the float upwards. The fluid has to flow through the constriction which is the gap between the float and the walls of the tube and so there is a pressure drop. Since the tube is tapered and the gap between the float and the tube walls increases as the float moves up the tube the pressure drop decreases as the float moves up the tube. The float moves up the tube until the fluid pressure is just sufficient to balance the weight of the float. The greater the flow rate the greater the pressure difference for a particular gap and so the float thus moves up the tube to a height which depends on the rate of flow. A scale alongside the tube can thus be calibrated to read

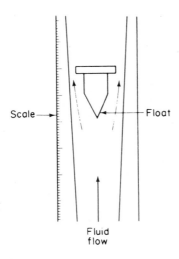

Figure 15.6 Rotameter

directly the flow rate corresponding to a particular height of float. The rotameter is cheap, reliable, has an accuracy of about $\pm 1\%$, and can be used to measure flow rates from about $30 \times 10^{-6}\,\text{m}^3/\text{s}$ to $1\,\text{m}^3/\text{s}$.

6 Variable orifice

The orifice flowmeter gives a non-linear relationship between the pressure difference and the flow rate. A linear relationship can be produced by means of a variable orifice. Figure 15.7 shows two forms,

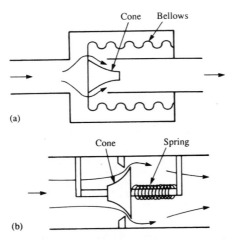

Figure 15.7 Variable orifice meter: (a) low flow rate (b) high flow rate

(a) for low flow rates and (b) for high flow rates. In both forms the fluid flow causes a cone to move axially and so change the orifice size. In one case the cone moves against the force exerted by a bellows and in the other that by a spring. The low flow rate version has a range up to about $0.05 \text{ m}^3/\text{s}$ and the high flow rate one to $3 \text{ m}^3/\text{s}$. They can be used with pressures up to $2 \times 10^7 \text{ Pa}$ and temperatures of $500\,°\text{C}$, often being used for steam.

Mechanical flowmeters

7 Target flowmeter

With the target flowmeter (Figure 15.8) the fluid impinges on a target, a disc, and it experiences a force. This force, via a force bar, causes movement of the flapper of a nozzle–flapper arrangement. The resulting change in pneumatic pressure is a measure of the force and also is used, by means of bellows, to restore the force bar and target to their undeflected position. The force is thus balanced through the force bar by the pneumatic pressure in the bellows. Target flowmeters have a range up to about $0.03 \text{ m}^3/\text{s}$ with an accuracy of $\pm 0.5\%$ and can be used for both liquids and gases. They can be used with viscous and dirty fluids.

The mass of fluid hitting the target per second is $\rho A v$, where ρ is the fluid density, A the target area and v the fluid velocity. This fluid has a momentum $(\rho A v)v$ and thus the change of momentum per second is $\rho A v^2$. The force acting on the disc will be the rate of change of momentum, hence

$$F = \rho A v^2$$

If d is the target diameter then

$$F = \rho(\pi d^2/4)v^2$$

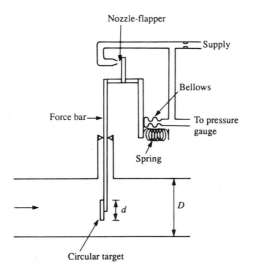

Figure 15.8 Target flowmeter

The fluid flowing past the target flows through an annular area of $\pi(D^2 - d^2)/4$, where D is the tube diameter. The quantity of fluid flowing through this space per second Q is thus

$$Q = [\pi(D^2 - d^2)/4]v$$

Hence

$$Q = \sqrt{(\pi/4)}[(D^2 - d^2)/d]\sqrt{(F/\rho)}$$
$$= C[D^2 - d^2)/d]\sqrt{(F/\rho)}$$

where C is a constant. Thus since the force is balanced by the pneumatic pressure in the bellows, the rate of flow is proportional to the square root of the pneumatic pressure.

8 Turbine flowmeters

The turbine flowmeter (Figure 15.9) consists of a multi-bladed rotor that is supported centrally in the pipe along which the flow occurs. The rotor rotates as a result of the fluid flow, the angular rate of rotation being proportional to the flow rate. The rate of revolution of the rotor can be determined using a magnetic pick-up. This could be a variable reluctance form of pick-up with the blades made of a ferromagnetic material and every time they pass the coil producing reluctance changes. The meter is expensive, offers some resistance to the fluid flow, is easily damaged by particles in the fluid, has good repeatability, an accuracy of about $\pm 0.3\%$ and a range up to about 1 m³/s. Turbine meters can be used for both liquids and gases.

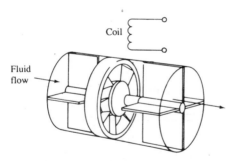

Figure 15.9 Turbine meter

9 Rotating vane

The fluid impinges tangentially on vanes attached to a rotor (Figure 15.10). The rotation of the rotor is monitored by a suitable pick-up or gearing to the shaft of the rotor. The meter can be used for low liquid flow rates. A similar system can be used for gases, being then referred to as an *anemometer*.

Positive displacement meters

This form of flowmeter divides up the flowing fluid into known volume packets and then counts them to give the total volume passing through the meter. If the volume delivered over a particular time is monitored then the volume rate of flow can be established. They are widely used for water meters, gas meters and fuel pump meters to determine the volume delivered.

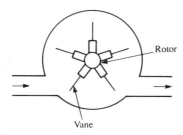

Figure 15.10   Rotating vane meter

10   *Rotating piston*

The rotating piston meter (Figure 15.11) consists of a cylindrical working chamber in which an offset hollow cylindrical piston rotates. Fluid is trapped by the rotating piston and swept round and out through the outlet. The number of rotations of the piston drive shaft is a measure of the volume that has passed through the meter. The meter is widely used for metering domestic water supplies. The accuracy is about $\pm 1\%$.

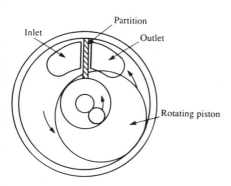

Figure 15.11   Rotating piston meter

11   *Reciprocating piston*

The reciprocating piston meter (Figure 15.12) has a piston which is driven by the incoming fluid which fills up the chamber as the piston is displaced to its maximum position. When this position is reached the slide valve reverses the side of the piston to which the fluid is admitted and so causes the piston to reverse its path. This drives the fluid already in the chamber out of the outlet. A ratchet attached to the piston rod is used to drive a counter. The meter is used with liquids, very accurate, $\pm 0.1\%$, and can operate over a very wide range.

12   *Nutating disc*

With the nutating disc (Figure 15.13) a disc acts like the piston in the reciprocating piston meter. The incoming fluid causes the disc to first

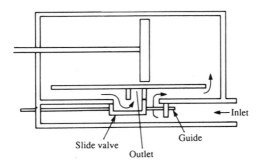

Figure 15.12 Reciprocating piston meter

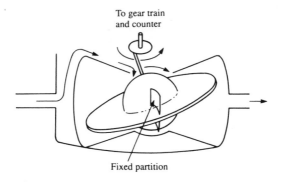

Figure 15.13 Nutating disc meter

move upwards at its right side and so fill the chamber. Then the incoming fluid enters the chamber at the other side and forces the left side of the disc upwards. This then results in the fluid already trapped in the chamber becoming expelled through the outlet. This movement up-and-down of the disc causes the spindle, protruding from the sphere on which it is mounted, to move in a circular path and drive a geared counter. The meter is used with liquids and has an accuracy of about $\pm 1\%$.

13 Rotating impeller
Figure 15.14 shows a version of the rotating impeller meter which is used for liquids, Figure 15.15 a version for gases. With both forms there are two rotors. The fluid flow causes them to rotate. Each time the rotors rotate they trap a volume of the fluid and move it from the inlet to the outlet. A counter is driven by a rotating rotor. The fluted rotor version is often used for oil flows up to about $1\,m^3/s$ and a pressure of 80 bar. The rotating lobe meter with gases has a range of about $0.003\,m^3/s$ to $3\,m^3/s$ with an accuracy of $\pm 1.0\%$.

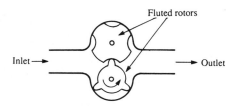

Figure 15.14 Fluted rotor meter

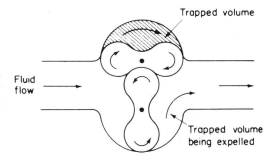

Figure 15.15 Rotating lobe meter

14 Rotating vane

The rotating vane meter (Figure 15.16) consists of a cylindrical rotor from which four retractable vanes protrude. The fluid flow against the vanes causes the rotor to rotate. As the rotor rotates the trapped fluid between vanes is swept round and out of the chamber. The number of revolutions of the rotor is thus a measure of the amount of fluid that has been passed through the meter. Accuracy is high, about $\pm 0.1\%$. The meter is widely used for measurements with oil and fuel.

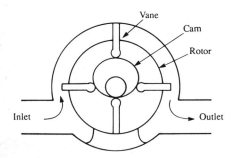

Figure 15.16 Rotating van meter

15 Diaphragm meter

Figure 15.17 shows one form of a diaphragm meter. It has four chambers, A, B, C and D which fill and empty with gas. For the situation shown in the figure A is emptying, B is filling, C is empty and D has just filled. This is then followed by the slide valve moving to the right to close the entrances to A and B and open those to C and D. Then A is empty, B full, C is filling and D emptying. The slide valve then moves further to the right to open A and B and close C and D. Then A fills, B empties, C is full and D empties. The slide valve then moves to the left to close A and B and open C and D. Then A is full, B empty, C empties and D fills. The sequence then repeats itself. Diaphragm meters are used for metering domestic gas supplies.

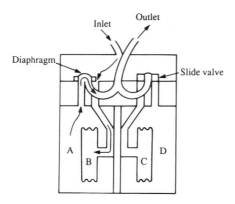

Figure 15.17 Diaphragm meter

16 Liquid sealed drum

The liquid sealed tube meter is used for the metering of gas. Gas enters the drum (Figure 15.18) by an inlet close to the drum centre. The gas is trapped between the shaped sheet and the liquid until the sheet emerges from the liquid. Then the trapped volume is discharged to the outlet.

Electromagnetic flowmeters

If a conductor of length L is moving with a velocity v at right angles to a magnetic field of flux density B (Figure 15.19) then an e.m.f. E is induced between the ends of the conductor, where

$$E = BLv$$

Thus if a conducting liquid is moving with an average velocity v through a tube which is at right angles to a magnetic field of flux density B then an e.m.f. E is induced and appears between two electrodes on opposite walls of the tube. The length L of the conductor is the diameter of the tube D, hence

$$E = BDv$$

Since the quantity rate of flow Q is $(\frac{1}{4}\pi D^2)v$ then

$$E = \frac{4BQ}{\pi D}$$

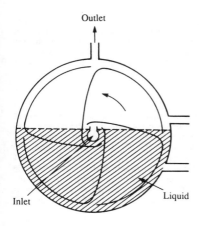

Figure 15.18 Liquid sealed drum

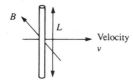

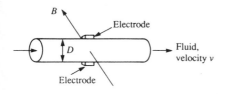

Figure 15.19 Conductor moving in magnetic field

17 Electromagnetic flowmeter

The magnetic field is produced by a pair of coils situated on either side of the fluid-carrying tube. If a constant direct current is used for these coils gas bubbles are formed round the electrodes (the effect being called polarization) and other electrochemical and thermochemical effects occur. These problems are overcome by using pulsed direct current (alternating current can lead to interference effects with the mains frequency), typically the direct current being switched on and off at a rate of about three times each second. During each cycle the induced e.m.f. signal is sampled five times. By using a microprocessor these signals can be analysed and a signal related to the flow, free of any zero errors, extracted.

Electromagnetic flowmeters can be used with a wide range of liquids provided the electrical conductivity is greater than about 1 μS/cm. It can be used with dirty liquids and slurries, being unaffected by changes in temperature, pressure, viscosity, density or conductivity. The velocity range is up to about 10 m/s with pipes of diameters between 32 mm and 1200 mm. Accuracy is about $\pm 1\%$.

Ultrasonic flowmeters

18 Doppler flowmeter

Figure 15.20 shows the principle of the Doppler flowmeter. A transmitter sends an ultrasonic wave of frequency f and velocity c into the fluid. It is reflected by bubbles, particles or eddies in the fluid which are moving with the fluid velocity v. The velocity of the ultrasonic waves relative to such particles is $(c + v \cos \theta)$ and thus the apparent frequency is $(c + v \cos \theta)f/c$. The particle then reflects the ultrasonic waves and acts as a transmitter moving with a velocity v relative to the receiver. The velocity of these waves relative to the receiver is $(c - v \cos \theta)$ and the apparent frequency $[c/(c - v \cos \theta)]$ $[c + v \cos \theta)f/c]$. Thus

$$\text{received frequency} = \frac{f(c + v \cos \theta)}{(c - v \cos \theta)}$$

v/c is relatively small and so second and higher powers of it can be neglected. The expression thus simplifies to

$$\text{received frequency} - \text{emitted frequency } f = \frac{2fv}{c}\cos \theta$$

The frequency difference is thus proportional to the fluid velocity and so volume rate of flow.

Ultrasonic flowmeters are relatively cheap. Accuracy, at the best, is normally only about $\pm 5\%$. Because of this the flowmeter is often used for just flow indication or as a flow switch.

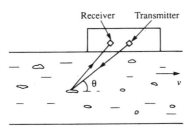

Figure 15.20 Doppler flowmeter

19 Time of flight flowmeter

This instrument consists of a pair of ultrasonic receiver–transmitters, one on each side of the pipe through which the fluid is flowing (Figure 15.21). The speed of an ultrasonic wave in one direction is $(c + v \cos \theta)$ and in the other direction $(c - v \cos \theta)$, where c is the speed in a stationary fluid. The time taken for an ultrasonic wave pulse to travel in one direction is thus $L/(c + v \cos \theta)$ and in the other direction $L/(c - v \cos \theta)$. If the receipt of a pulse is used to trigger the transmission of the next one then the frequency with which the pulses

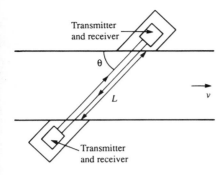

Figure 15.21 Time of flight flowmeter

are emitted in each direction is $(c + v\cos\theta)/L$ and $(c - v\cos\theta)/L$. The difference in these frequencies is thus

$$\text{frequency difference} = \frac{(c + v\cos\theta)}{L} - \frac{(c - v\cos\theta)}{L}$$

$$= \frac{2v\cos\theta}{L}$$

The frequency difference can thus be used as a measure of the fluid velocity.

The time of flight meter can be used for pipes from 75 mm diameter to 1500 mm diameter, fluid velocities between about 0.2 m/s and 12 m/s with an accuracy of about $\pm 1\%$ or better. It can also be used for open-channel and river flow.

20 Cross-correlation flowmeter

This method assumes that in the flowing fluid there are some random fluctuations, such as turbulence, bubbles or particles, which can be detected. Figure 15.22 shows the form of the instrument when the means of detecting the fluctuations is ultrasonics. Any fluctuation passing between a transmitter and receiver affect the received signal, changing its amplitude and phase. The signals received by the two receivers are, after amplification and filtering, fed to the correlator. This evaluates the cross-correlation function for the two signals. The correlation function is a maximum when there is maximum similarity between the two signals. The time difference t between the reception of these two highly correlated signals is then determined and so the flow velocity v obtained, $v = L/v$. A problem with such systems has been the slow response time.

Oscillatory

21 Vortex flowmeter

When a fluid flow encounters a body, the layers of the fluid close to the surfaces of the body are slowed down. With a streamlined body these boundary layers follow the contours of the body until virtually meeting at the rear of the object. This results in very little wake being produced. With a non-streamlined body, a so-called bluff body, the boundary layers detach from the body much earlier and a large wake is produced. When the boundary layer leaves the body surface it rolls up

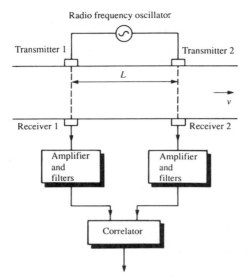

Radio frequency oscillator

Figure 15.22 Cross-correlation flowmeter

into vortices. These are produced alternatively from the top and bottom surfaces of the body (Figure 15.23). The result is two parallel rows of vortices moving downstream with the distance between vortices in each row being the same, a vortex in one row occurring half way between those in the other row. The number of vortices produced per second f from each surface of the bluff body is

$$f = \frac{Sv_b}{d}$$

where v_b is the mean fluid velocity at the bluff body, d the width of the body and S is a quantity, virtually a constant, called the *Strouhal number*. At the body the fluid flows through an area of $\pi D^2/4 - Dd$, where D is the tube diameter. The body is assumed to present a rectangular surface of width d and extending right across the diameter of the tube and so an effective area of about Dd. The velocity v_b at this point is related to the velocity v some distance from the body and the rate of volume flow Q by

$$Q = (\pi D^2/4)v = (\pi D^2/4 - Dd)v_b$$

Hence

$$f = \frac{SD^2v}{\pi D^2 - Dd} = \frac{SD^2Q}{(\pi D^2/4)(\pi D^2/4 - Dd)}$$

$$= \frac{4SQ}{\pi D^2 d[1 - (4d/\pi D)]}$$

Because a variety of different shape bluff bodies (Figure 15.24) are used the equation is modified by introducing a bluff body coefficient k which has different values for different shapes.

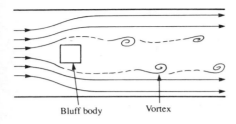

Figure 15.23 Vortex shedding

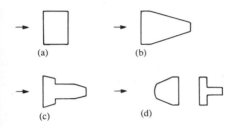

Figure 15.24 Bluff body shapes

$$f = \frac{4SQ}{\pi D^2 d[1 - (4dk/\pi D)]}$$

Thus for a particular bluff body the frequency is proportional to the flow rate.

A number of methods are available for the measurement of the frequency (Figure 15.25). A thermistor might be located in the upstream surface of the bluff body. The thermistor, heated as a result of a current passing through it, senses vortices due to the cooling effect caused by their breaking away producing a resistance change. Another method has the vortices passing through a beam of ultrasonic waves. The resulting amplitude changes to that wave can be monitored. Another form of detector uses a piezo-electric crystal mounted in the bluff body. Flexible diaphragms react to the pressure disturbances produced by the vortices and are detected by the piezo-electric crystal. In one form of this system the piezo-electric sensor is mounted in a second bluff body mounted downstream of the first (as in Figure 15.24(d)).

Vortex flowmeters are used for both liquids and gases, having an output which is independent of density, temperature or pressure, and have an accuracy of about $\pm 1\%$. They are used at pressures up to about 10 MPa and temperatures of 200 °C.

22 Swirl meter

With the swirl meter (Figure 15.26) the fluid is made to swirl or spin by passing through curved blades. The oscillations of this swirling liquid are detected by a temperature sensor. This is a thermistor which is

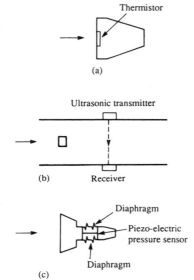

Figure 15.25 Detection systems

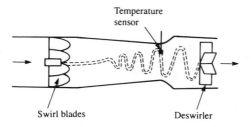

Figure 15.26 Swirl meter

heated as a result of carrying an electrical current. The heat lost by the thermistor is affected by whether it is in a swirl or not. The result is that the temperature of the thermistor, and hence its resistance, oscillates at the same frequency as the swirl. This frequency is proportional to the quantity rate of flow. Swirl meters are used with liquids in the range $6 \times 10^{-4}\,\text{m}^3/\text{s}$ to $2\,\text{m}^3/\text{s}$, with gases $10^{-3}\,\text{m}^3/\text{s}$ to $3\,\text{m}^3/\text{s}$. Accuracy is about $\pm 1\%$.

Direct mass flowmeters

23 Coriolis mass flowmeter

A body of mass M moving with constant linear velocity v and subject to an angular velocity ω experiences an inertial force at right angles to

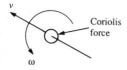

Figure 15.27 Coriolis force

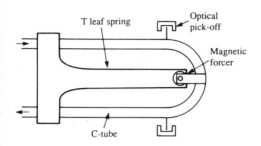

Figure 15.28 Coriolis flowmeter

the direction of motion, this being known as a *Coriolis force* (Figure 15.27).

Coriolis force $= 2M\omega v$

The Coriolis mass flowmeter basically consists of a C-shaped pipe (Figure 15.28) through which the fluid flows. The pipe, and fluid, is given an angular acceleration by being set into vibration, this being done by means of a magnet mounted on the C in a coil mounted on the end of a spring. Oscillations of the spring then set the C-tube into oscillation. The angular acceleration thus alternates in direction. At some instant the Coriolis force acting on the fluid in the upper limb is in one direction and in the lower limb in the opposite direction, this being because the velocity of the fluid is in opposite directions in the upper and lower limbs. The resulting Coriolis forces on the fluid in the two limbs are thus in opposite directions and cause the limbs to become displaced. When the direction of the angular velocity is reversed then the forces reverse in direction and the limbs become displaced in the opposite direction. These displacements are proportional to the mass flow rate of fluid through the tube. The displacements are monitored by means of optical transducers, their outputs being a pulse with a width proportional to the mass flow rate. The Coriolis flowmeter can be used for liquids or gases and gives an accuracy of $\pm 0.5\%$. It is unaffected by changes in temperature or pressure.

Further reading: For analysis of the Coriolis force see advanced texts on dynamics, e.g. Hibbeler, R. C. (1985), *Mechanics for Engineers*, Macmillan; for details of the flowmeter see Loxton, R. and Pope, P. (eds) (1986), *Instrumentation: A Reader*, Open University Press.

24 *Thermal mass flowmeter*

This flowmeter consists of two temperature sensors mounted with one upstream and the other downstream of a heater (Figure 15.29). The difference in temperature between the sensors depends on the rate of mass flow. The two sensors are mounted in adjacent arms of a Wheatstone bridge. The out-of-balance potential difference from the bridge is thus a measure of the temperature difference and hence the rate of mass flow. Such a meter is used for gas flows in the range 2.5×10^{-10}g/s to 5×10^{-3}kg/s with an accuracy of $\pm 1\%$.

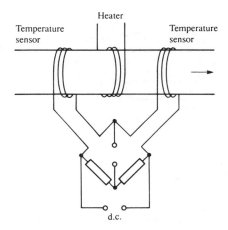

Figure 15.29 Thermal mass flowmeter

Inferential mass flowmeters

Inferential measurements are based on determining the volume flow rate and the fluid density as separate measurements and then computing from those results the mass flow rate. In the case of a pure liquid, since the density depends only on the temperature, if the temperature is reasonably constant then the density may be assumed to be constant and thus a determination of just the volume flow rate gives a measure of the mass flow rate. Where gases and non-homogeneous liquids are concerned both the volume rate and density need to be measured.

25 *Turbine-vibrating element*

A turbine flowmeter (see earlier this chapter) is used to give a frequency output signal proportional to the quantity rate of flow. A vibrating element density transducer (see Chapter 12) is used to give a frequency output signal related to the fluid density. These two signals can be combined by a computer to give an output of the mass flow rate. These methods can be used for liquids and gases.

Pressure point velocity measurements

26 *Pitot static tube*

With the Pitot static tube (Figure 15.30) the pressure difference is measured between a point in the fluid where the fluid is in full flow, the

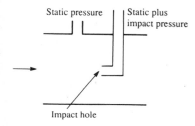

Static pressure | Static plus impact pressure

Impact hole

Figure 15.30 Pitot static tube

impact pressure P_I, and a point at rest in the fluid, the static pressure P_s. The difference in pressure is thus due to the kinetic energy of the fluid and so applying the conservation of energy (Bernouilli's law, see earlier this chapter),

$$\frac{P_I}{\rho} = \frac{P_s}{\rho} + \frac{v^2}{2}$$

where ρ is the fluid density and v its velocity at the impact point. Hence

$$v = \sqrt{\left[\frac{2(P_I - P_s)}{\rho}\right]}$$

This is the relationship for an incompressible fluid such as a liquid, for a compressible fluid such as a gas the relationship needs modification. This is because the fluid density at the impact ρ_I and static ρ_s holes will not be the same. The equation then becomes

$$\frac{\gamma}{\gamma - 1}\frac{P_I}{\rho_I} = \frac{\gamma}{\gamma - 1}\frac{P_s}{\rho_s} + \frac{v^2}{2}$$

Assuming that the density changes are adiabatic

$$\frac{P_I}{\rho_I{}^\gamma} = \frac{P_s}{\rho_s{}^\gamma}$$

Hence

$$v = \sqrt{\left[2\frac{\rho}{\gamma - 1}\frac{P_s}{\rho_s}\left\{\left(\frac{P_I}{P_s}\right)^{(\gamma - 1)/\gamma} - 1\right\}\right]}$$

For air the difference in densities between the static and impact holes is negligible for speeds less than 100 m/s and so the incompressible equation can be used.

The pressure difference is often measured with a diaphragm form of pressure gauge. They can be used for both liquids and gases and have an accuracy of about ± 1 or 2%. A problem with the Pitot static tube is that the holes can easily become blocked hence it is mainly used with gases rather than liquids.

27 Annubar tube
This is a form of Pitot tube, with four impact holes in a bar across the width of the tube (Figure 15.31). The spacing of the holes is such that each responds to the pressure of equal annuli segments of the flow. The average of the pressures from these four holes is then indicated by an inner tube. The static pressure is obtained from a probe facing downstream. This arrangement gives an accuracy of $\pm 1\%$ or better.

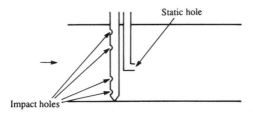

Figure 15.31 Annubar tube

Thermal point velocity measurements

28 Hot wire anemometer

The hot wire anemometer consists of a small resistance wire element mounted in the flow (Figure 15.32). An electric current through the wire causes its temperature to rise to a value which is determined by the rate at which it loses heat. This depends on the velocity of the fluid. At equilibrium

$$i^2 R = hA(T_s - T_f)$$

where i is the sensor current, R its resistance, A its effective area, T_s its temperature and T_f the temperature of the fluid. The heat transfer coefficient h depends on the fluid velocity v,

$$h = C_0 + C_1 \sqrt{v}$$

where C_0 and C_1 are constants. Thus

$$i^2 R = A(C_0 + C_1 \sqrt{v})(T_s - T_f)$$

Usually the resistance, and hence the temperature, of the element is kept constant by changing the current. The current then becomes a measure of the fluid velocity.

$$i^2 = C_2 + C_3 \sqrt{v}$$

where C_2 and C_3 are constants. The hot wire anemometer is used for gas velocities from 0.1 m/s to 500 m/s at temperatures up to 750 °C, and for liquids from 0.01 m/s to 5 m/s. A sensor made using a thin film wrapped round a cylinder rather than a wire can be used with liquids from 0.01 m/s to 25 m/s. Accuracies are about $\pm 1\%$.

Calibration

Commonly used methods for the calibration of flowmeters are: for liquids or gases, measurement of the mass passed through a pipe in a measured time, or with liquids to directly measure the volume passed through a pipe in a measured time, or for liquids or gases to calibrate against a standard flowmeter.

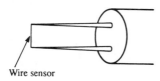

Wire sensor

Figure 15.32 Hot wire anemometer

16 Force

The unit of force is the *newton* and is that force which when applied to a mass of one kilogram gives it an acceleration of $1 \, \text{m/s}^2$. The *weight* of a body of mass m at rest relative to the surface of the earth is the force exerted on it due to gravity and is equal to mg, where g is the acceleration due to gravity. Since weight is a force it has the unit of newton, however weight measurements are often specified in terms of kilograms, this being the mass that would give rise to the weight force at the local value of the acceleration due to gravity. Table 16.1 lists the force, and weight, measurement systems discussed in this chapter and gives their main characteristics.

Further reading: Noltingk, B. E. (1985), *Jones' Instrument Technology*, vol. 1 (*Mechanical Measurements*), Butterworth-Heinemann.

Table 16.1 Force measurement systems

Principle	System	Characteristics
Lever-balance	1 Equal arm balance	Weights up to 1000 kg, very accurate.
	2 Unequal arm balance	Bulky, very accurate.
Force-balance	3 Force-balance	Very accurate, high stability, range 0.1 N to 1 kN, static and dynamic force measurements.
Elastic elements	4 Spring balance	Low accuracy, cheap, range 0.1 N to 10 kN, static forces only.
	5 Proving ring	Accuracy ±0.2 to 0.5%, range 2 kN to 2 MN, static forces only.
	6 Load cells	Accuracy ±0.01 to 1%, range 5 N to 40 MN, static and dynamic force measurements.
	7 Piezo-electric	Accuracy ±0.5 to 1.5%, range 5 kN to 1 MN, dynamic forces only.
Pressure	8 Hydraulic pressure	Accuracy ±0.25% to 1%, range 5 kN to 5 MN, static and dynamic force measurements.

Lever-balance methods

Lever-balance methods depend on the principle of moments, i.e. at static equilibrium the algebraic sum of the clockwise moments about an axis must equal the anticlockwise moments. The moment of a force about an axis is the product of the force and the perpendicular distance from its line of action to the axis.

1 Equal arm balance

Figure 16.1 shows the basic principle of an equal arm balance. It consists of a rigid beam pivoted on a knife edge with the unknown force F_u being applied at an equal distance from the pivot axis to the

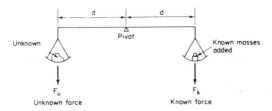

Figure 16.1 Equal arm balance

balancing, known, force F_k. At equilibrium when the beam is horizontal

$$F_u d = F_k d$$

where d is the distance from the lines of action of the forces to the pivot axis. Thus $F_u = F_k$. The forces usually compared by this method are weights. Thus $m_u g = m_k g$ and since the acceleration due to gravity is the same at both ends of the beam then the unknown mass m_u equals the known mass m_k. The known mass is adjusted until this equilibrium occurs. This type of balance is mainly used for weighing chemicals and versions can be used for weights up to 1000 kg with very high accuracy.

2 Unequal arm balance

Figure 16.2 shows the basic principle of an unequal arm balance. The unknown force F_u is balanced by moving a constant known force F_k to different distances from the pivot axis. At equilibrium when the bar is horizontal

$$F_u a = F_k b$$

Since a is a constant then F_u is proportional to the distance b. A graduated scale along the beam enables the weight to be directly read when balance occurs. The range of the balance is changed by adding masses to the end of the beam, a distance c from the pivot axis. Such a form of balance is widely used in industry. It is bulky but can be very accurate.

Force-balance methods

Force-balance methods depend on the unknown force causing a displacement which is monitored by a displacement transducer. This

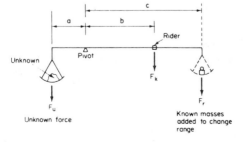

Figure 16.2 Unequal arm balance

gives a signal which, after amplification, is used to activate another transducer to exert a force which just balances the unknown force and so returns the displacement to zero.

3 Force-balance

Figure 16.3 shows the basis of one form of force-balance. The unknown force causes the ferromagnetic core of a LVDT (see item 12, Chapter 8) to move. The electrical output from the LVDT is then amplified before passing through a coil in a magnetic field. A consequence of this is that a magnetic force acts on the coil and hence on the member along which the unknown force acts. When the magnetic force balances the unknown force the amplified LVDT signal is a measure of the unknown force. Such methods of measuring forces have high stability, very high accuracy, a range from about 0.1 N to 1 kN and can be used for both static and dynamic force measurements.

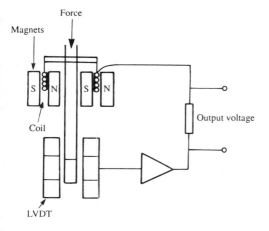

Figure 16.3 A force-balance

Elastic elements

The basic principle involved in force measurement systems with elastic elements is that the change in length of such an element is directly proportional to the applied force, i.e. Hooke's law is obeyed. The change in length is generally converted into some other variable by another transducer.

4 Spring balance

With the spring balance (Figure 16.4) the extension of a spring is used as a measure of the applied force. Spring balances have low accuracy as the extensions produced are relatively small, are only suitable for static force measurement, have ranges within about 0.1 N to 10 kN, and are cheap.

5 Proving ring

With proving rings (see item 20, Chapter 8 and Figure 8.22) forces are applied across a diameter of a ring and cause it to distort, the amount of distortion being proportional to the force. The distortion can be measured with a dial test indicator gauge. For more accurate

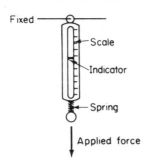

Figure 16.4 Spring balance

measurements micrometer screws or electrical transducers such as an LVDT can be used. Proving rings can thus be direct reading or give a display at a distance. They can be very accurate, ± 0.2 to 0.5%, have a range of the order of $2\,kN$ to $2000\,kN$ and are only used for static force measurements.

6 Load cells

Strain gauges may be used to determine the deformation of some elastic member when subject to force and so provide a measurement of the force. Such a system is known as a load cell. The elastic member might be a hollow or solid cylinder (see item 21, Chapter 8 and Figure 8.23), a ring, a cantilever, a shear member, or a diaphragm (see Figure 18.7). Generally four strain gauges are used and when forces are applied two of the gauges are in tension and the other two in compression. The gauges form the arms of a Wheatstone bridge with opposite arms being gauges subject to tension. The use of four identical strain gauges, one in each arm of the bridge, eliminates the effect of temperature on gauge resistance since any temperature change affects equally each arm of the bridge and so produces no out-of-balance potential difference. When there is no load all the four gauges are the same resistance and so the output potential difference from the bridge is zero. When subject to a force an out-of-balance potential difference is produced which is related to the size of the force. Load cells give a display at a distance, have a rapid response, can be used for both static and dynamic forces, are robust, have an accuracy of about $\pm 0.01\%$ to 1.0%, and have a range within about $5\,N$ to $40\,MN$, depending on the form of the load cell.

7 Piezo-electric methods

Compressive forces applied to opposite faces of certain crystals in causing a contraction result in the faces becoming charged, the amount of the charge being proportional to the size of the force (see item 15, Chapter 8). This is known as the *piezo-electric effect*. It is used for dynamic forces, not being suitable for the measurement of static forces. It is small, robust, used for compressive forces in the range $5\,kN$ to $1\,MN$, and has an accuracy of about ± 0.5 to 1.5%.

Pressure methods

8 Hydraulic pressure method

A *hydraulic pressure force measurement system* (Figure 16.5) consists of a chamber containing oil connected to a pressure gauge, possibly a Bourdon tube gauge. The chamber has a diaphragm to which the force

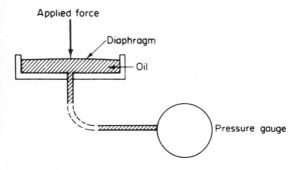

Figure 16.5 Hydraulic pressure force measurement system

is applied. This leads to a change in pressure of the oil, which then shows up as a change in reading of the pressure gauge. The gauge can be calibrated directly in terms of the applied force. Such an instrument has the advantage of not requiring any electrical power, can give a local or distant reading, gives a relatively rapid response to force changes, can be used for both static and dynamic forces, can be used within the range 5 kN to 5 MN, and has an accuracy of about $\pm 0.25\%$ to 1.0%.

17 Level

Table 17.1 lists the methods for the determination of level discussed in this chapter and their general characteristics. Where the method involves the transducer being in the liquid in the main vessel then a *stilling chamber* may be required. This is just a tube which surrounds the transducer, e.g. a float, in the liquid and fills to the same level as the liquid in the container by means of small holes in its sides and so prevents disturbances directly affecting the transducer. Such a chamber can however produce time lags. Frequently the transducer is used in a side arm to the vessel. In such situations problems can arise due to temperature differences between the liquid in the vessel and that in the side arm.

Further reading: Cho, C. H. (1982), *Measurement and Control of Liquid Level*, Instrument Society of America; Noltingk, B. E. (ed.) (1985), *Jones' Instrument Technology*, vol. 1 (*Mechanical Measurements*), Butterworth-Heinemann.

Table 17.1 Level measurement methods

Principle	System	Characteristics
Sight	1 Dipstick	Simple, cheap, non-continuous reading method for liquids.
	2 Hook gauge	Simple, cheap method for small changes in liquid level.
	3 Sight glasses	Cheap, can be used for open and pressure vessels, up to 700 MPa and 300 °C.
Floats	4 Tape-float gauge	Cheap, can be used with corrosive liquids and over a wide temperature range.
	5 Potentiometer float gauge	Voltage output, cheap, can be used for corrosive liquids and over a wide temperature range.
	6 Magnetic float gauge	Can be used with corrosive liquids and over a wide temperature range.
Displacers	7 Spring balance gauge	Can be used for liquid interfaces, accurate, and has a limited range.
	8 Torque tube gauge	Can be used for liquid interfaces, accurate, gives electrical or pneumatic output, and has a limited range.
Pressure	9 Pressure gauge	Can be used for open and pressure vessels, form of output depends on type of pressure gauge used.

Table 17.1 (continued)

Principle	System	Characteristics
	10 Bubbler method	Can be used with corrosive liquids and slurries, form of output depends on type of pressure gauge used.
Weight	11 Load cells	Can be used for liquids, slurries, solids and corrosive liquids. Electrical output from strain gauges.
Electrical	12 Conductivity level indicator	Only indicates when critical level reached, restricted to high conductivity liquids.
	13 Resistance gauge	Can be used for solids or liquids.
	14 Capacitive gauge	Can be used with corrosive liquids and at high temperatures and pressures. Rugged and accurate.
Ultrasonics	15 Echo type	Can be used for solids and liquids, particularly for corrosive liquids, interfaces, relatively expensive.
Radiation	16 Absorption type	Can be used for solids and liquids, particularly for corrosive or high temperature situations. Relatively expensive.
Thermal	17 Hot wire element	Detects when a critical level has been reached.
	18 Thermistor	Detects when a critical level has been reached.

Sight methods

Sight methods depend on the visual observation of the level against some scale. With such methods time is required to make the reading but accuracy of the order of $\pm 1\%$ is possible.

1 Dipstick

The dipstick (Figure 17.1) is a simple, cheap, method of determining the level of a liquid in a container, e.g. the oil level in a car engine. It is just a stick which is held vertically, from some constant position, in the liquid. The stick is then removed and the mark left by the liquid on the stick enables the position of the level to be determined. Dipsticks have the disadvantage of needing to be removed and examined before a level reading can be obtained.

2 Hook gauge

The hook gauge is a form of dipstick. Figure 17.2 shows one form. The position of the tip of the hook is adjusted by the screw until it just touches the surface of the liquid. The amount by which the screw is rotated thus gives a measure of the level.

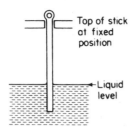

Figure 17.1 Dipstick

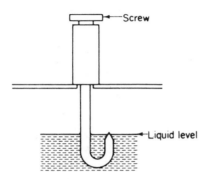

Figure 17.2 Hook gauge

3 Sight glass

With sight glasses the level of the liquid is directly observed against a scale. Figure 17.3 shows two forms of tubular sight glass, one being for use with liquid in an open tank and the other for a pressure vessel. The gauges are transparent tubes made of glass or plastic. They can be used at pressures up to about 3 MPa and temperatures up to about 200 °C. A flat form of sight glass (Figure 17.4) can be used for pressures as high as 700 MPa and temperatures of 300 °C. The main disadvantage of sight glasses is that they have to be read where the tank is located. The accuracy of the readings depends on the cleanliness of the glass and the liquid.

Float methods

Float methods involve a float resting in the surface of the liquid and moving up or down as the liquid level changes. The movement of the float is transmitted to a pointer moving over a scale by a variety of mechanisms. Errors are produced by the build-up of sediments on floats and corrosion affecting the mass of floats.

4 Tape-float gauge

With the tape-float gauge (Figure 17.5) the float is attached to one end of a metal tape which passes over a pulley to a counterweight. When the liquid level changes the float moves, the counterweight keeping the

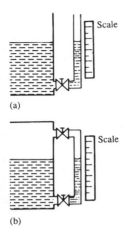

(a)

(b)

Figure 17.3 Tubular sight glasses (a) open tank (b) pressure vessel

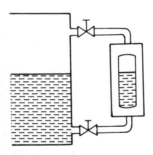

Figure 17.4 Flat sight glass

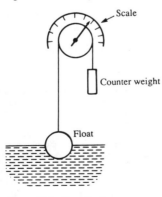

Figure 17.5 Tape-float gauge

tape taut. This requires movement of the tape which in turn causes the pulley wheel to rotate and hence a pointer moves over a scale. The choice of float material depends on the corrosive nature of the liquid, the most commonly used being brass, copper, nickel alloys or stainless steel. The float is generally a sphere or a cylinder. The gauge can be used over a large temperature range, with corrosive liquids, but is limited to essentially open vessels.

5 Potentiometer float gauge

A voltage output is obtained with the potentiometer float gauge (Figure 17.6). The float is at one end of a pivoted rod with the other end connected to the slider of a potentiometer. Changes in level cause the float to move, hence moving the potentiometer slider over the resistance track and so giving a potential difference output related to the level. The arrangement is widely used for the determination of the level, and hence amount, of fuel in motor car fuel tanks.

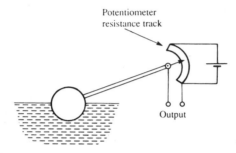

Figure 17.6 Potentiometer float gauge

6 Magnetic float gauge

The magnetic float gauge has a doughnut shaped magnetic material float which slides up and down round a sealed tube with the liquid level and causes a magnet inside the tube to move up and down in sympathy with it (Figure 17.7). This magnet is at the end of a lever and its motion up and down results in a pointer moving across a scale. The advantages of this are that the 'float' arrangement inside the tube does not come into contact with the liquid and there is no need for leakproof seals around the float shaft. Hence the arrangement is particularly useful for corrosive fluids.

Displacer gauges

When an object is partially or wholly immersed in a fluid it experiences an upthrust force equal to the weight of fluid displaced by the object (*Archimedes' principle*). Thus a change in the amount of an object below the surface of a liquid will result in a change in the upthrust. The net force acting downwards on such an object is its weight minus the upthrust and thus depends on the depth to which the object is immersed in the fluid, i.e. the amount of liquid it is displacing.

For a vertical cylinder of cross-sectional area A in a liquid of density ρ, if a height h of the cylinder is below the surface then the upthrust is $hA\rho g$ and so the apparent weight is $(mg - hA\rho g)$, where m is the mass of the cylinder and g the acceleration due to gravity. Displacer gauges

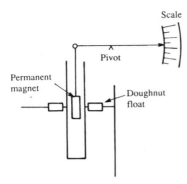

Figure 17.7 Magnetic float gauge

need calibrating for particular liquids since the upthrust depends on the density of the liquid concerned. They can be used to determine the interface between two liquids by virtue of their different densities.

7 *Spring balance gauge*

This just consists of a displacer, a cylinder, suspended from a spring balance (Figure 17.8). As the level of liquid rises so the amount of the displacer below the liquid surface increases and the upthrust increases. The result is that the apparent weight of the displacer decreases as the level increases.

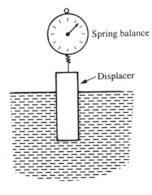

Figure 17.8 Spring balance displacer gauge

8 *Torque tube gauge*

The torque tube level gauge (Figure 17.9) has a displacer which pushes upwards against a rod and causes twisting of a tube. The twist of the tube can be monitored by means of strain gauges which thus give an electrical response related to the upthrust and hence level of the liquid. Another version has the twisting of the tube moving the flapper of a

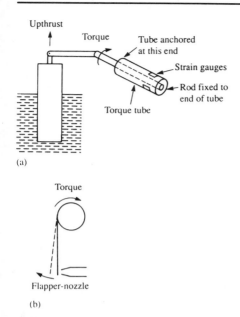

Figure 17.9 Torque tube level gauge (a) strain gauge form (b) pneumatic adaptation

flapper–nozzle arrangement (see item 25, Chapter 8) and so gives a pneumatic output.

Pressure methods

The pressure due to a height h of liquid above some level is $h\rho g$, where ρ is the density of the liquid and g the acceleration due to gravity.

9 Pressure gauge method

A measurement of the pressure difference between the surface of the liquid and the bottom of the liquid will give a measure of the level of the liquid in the tank. If the tank is open to the atmosphere (Figure 17.10(a)) then all that is necessary is to measure the gauge pressure, i.e. the pressure difference from atmospheric pressure. With a closed tank the pressure difference has to be measured between the bottom of the liquid and the gases above the liquid surface (Figure 17.10(b)). The pressure gauges used are generally diaphragm instruments (see Chapter 18) and have to be located at the same level as the minimum liquid level in the tank. If the zero level position differs from that of the pressure gauge then a correction has to be made, e.g. if the gauge is below the zero level then a pressure due to that height difference must be subtracted from the pressure readings.

10 Bubbler method

The bubbler method uses a pipe which dips to virtually the bottom of the tank (Figure 17.11). The pipe is connected to a constant pressure air, or other suitable gas, supply and air bubbles out of the bottom of

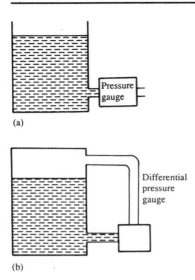

(a)

(b)

Figure 17.10 Pressure type of level gauge (a) open vessel (b) closed vessel

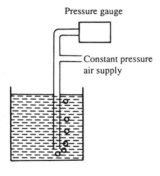

Figure 17.11 Bubbler method

the pipe. The greater the level of the liquid above the bottom of the pipe the greater the pressure that is needed before the bubbles can emerge. The escaping gas thus limits the pressure in the equipment which is then proportional to the height of the liquid above the bottom of the tube. The pressure can be measured by means of a diaphragm pressure gauge (see Chapter 18). A problem with this method is that the gas introduced into the liquid can affect processes involving the liquid, however it does have the advantage of only having the tube in the liquid and so with a non-corrosive material for the tube can be employed with corrosive liquids. It can also be used with slurries.

Weight methods

The weight of a liquid in a container is given by the product of its density and the liquid volume. If the container has a constant cross-sectional area then the volume is proportional to the height of the liquid and so the weight is directly related to the liquid level.

11 Load cells

Load cells (see item 21, Chapter 8) are included in the supports for the container and hence give a response related to the weight of the container and its contents. Since the weight depends on the level of liquid in the container then the load cells give responses related to liquid level. Since the load cells are completely isolated from the liquid the method is useful for corrosive liquids. Problems can occur if the tanks containing the liquid are subject to sideways forces, e.g. from the wind. The method can be used with liquids, slurries and solids.

Electrical methods

12 Conductivity level indicator

Conductivity methods can be used to indicate when the level of a high electrical conductivity liquid in a container reaches a critical level. One form has two probes, one probe mounted in the liquid and the other either horizontally at the required level or vertically with its lower end at the critical level. When the liquid is short of the required level the resistance between the two probes is high since part of the path between the probes is air. However, when the liquid level reaches the level of the air probe the resistance between the probes drops, this drop in resistance enabling a significant current to flow in an electrical circuit and so indicate that the critical level has been attained. An array of vertical probes might be used, each having its lower end at a different depth. Such an arrangement can be used to monitor variations in level. Foaming, splashing and turbulence can affect the result. One version of conductivity gauges does however use this fact to make determinations on the amount of froth on a liquid surface. It has an array of vertical probes. Resistances changes are produced when the froth reaches a probe and when the liquid does, the changes in resistance when these events occur being different for the froth and the liquid and hence capable of being distinguished.

13 Resistance level gauge

One form of resistance level gauge has a resistance element in the form of a strip (Figure 17.12). The strip has a conductive base strip which has close to it a flat resistance element, the entire element being in a protective, electrical insulating, sheath. When the element is vertically in a liquid the pressure acting on that part below the surface forces the

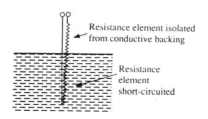

Figure 17.12 Resistance level gauge

resistance element into close contact with the conductive strip and short circuits it. The total resistance of the element thus depends on how much of it is below the liquid surface. The method can be used for solids or liquids.

14 Capacitive level gauge

A common form capacitive level gauge consists of two concentric cylinders, or a circular rod inside a cylinder, with the liquid between them (Figure 17.13). If the vessel containing the liquid is metal it might act as one plate of the capacitor, with a metal rod in the liquid acting as the other. If the liquid is an electrical insulator then the capacitor plates can be bare metal, if the liquid is conducting then they are coated with an insulator, e.g. Teflon. The arrangement consists essentially of two capacitors in parallel, one formed between the plates inside the liquid and the other from that part of the plates in the air above the liquid. A change in the liquid level changes the total capacitance of the arrangement (see item 8, Chapter 8). Errors can arise as a result of temperature changes since such a change will produce a change in capacitance without any change in level occurring, and also if the electrodes become coated with materials from the liquid. The method can be used for pressures up to about 14 MPa and temperatures of 350 °C and, by suitable choice of materials for the cylinders, for corrosive liquids. The probe is rugged and the method capable of accuracy.

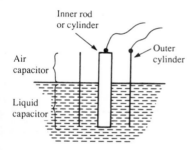

Figure 17.13 Capacitive level gauge

Ultrasonic methods

Ultrasonic waves are pressure waves with frequencies greater than about 20 kHz. The speed of such waves depends on the elasticity and density of the medium through which they are travelling and is affected by changes in temperature. When ultrasonic waves are incident on an interface, some of the wave energy passes through the interface and some is reflected. For the reflected wave the angle of incidence equals the angle of reflection.

15 Echo type

In one version, an ultrasonic transmitter/receiver is placed above the surface of the liquid (Figure 17.14). Ultrasonic pulses are produced, travel down to the liquid surface and are then reflected back to the receiver. The time taken from emission to reception of the reflected pulse can be measured. The time taken depends on the distance of the liquid surface from the transmitter/receiver and so the position of the level of the liquid can be determined.

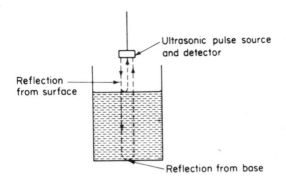

Figure 17.14 Ultrasonic level gauge

Another version involves placing the transmitter/receiver in the liquid at the bottom of the tank. The ultrasonic pulses are then directed vertically upwards and reflected from the liquid surface back down to the receiver. The time from emission to reception of the pulses is measured.

Ultrasonic methods can be used for solids, liquids, liquid–liquid interfaces and liquid–solid interfaces. Because the receiver/transmitter can be mounted outside the liquid it is particularly useful for corrosive liquids. Errors are produced by changes in temperature since they affect the speed of the sound wave. Such errors are typically about 0.18% per °C.

Nuclear radiation methods

Gamma radiation is short wavelength electromagnetic waves emitted by radioactive sources. It is used for level measurements in preference to other nuclear radiations because of its penetrating power. It is absorbed on passing through a medium, the amount of absorption depending on the length of the path in the medium and its density.

$$I = I_0 e^{-\mu x}$$

where I_0 is the incident radiation level, I the level after traversing a thickness x of a medium with an absorption coefficient μ. This coefficient depends on the density of the material.

Further reading: Loxton, R. and Pope, P. (eds) (1986), *Instrumentation: A Reader*, Open University.

16 Absorption methods

Gamma radiation is emitted from a radioactive source, generally cobalt-60, cesium-137 or radium-226, and passes through the container walls and any intervening liquid before reaching a nuclear radiation detector (see Chapter 19). The intensity of the detected radiation depends on the amount of liquid between the source and the detector. Figure 17.15 shows two possible arrangements. With a compact source and extended detector, level changes over the length of the detector can be detected. With a compact source and a compact detector the output from the detector is quite sensitive to changes in level over only a small range. Such methods can be used for liquids, slurries and solids, and, since no elements of the system are in the liquid, for corrosive and high temperature liquids. The equipment is

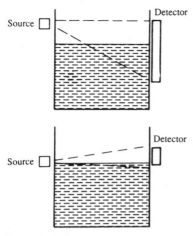

Figure 17.15 Radiation level measurement systems

relatively expensive and there is the problem of the hazardous nature of the radioactive source.

Thermal methods

17 Hot wire element

A hot wire element can be used to detect the surface of a liquid and so determine when a critical level has been reached (Figure 17.16). The rate at which heat is conducted away from a current-carrying wire depends on the medium in which the wire is located (see item 2, Chapter 8). Thus the temperature of such a wire depends on the medium. Hence, since the electrical resistance of the wire depends on its temperature, there will be a change in the resistance when the wire passes from above the liquid surface to below it. If the wire constitutes one arm of a Wheatstone bridge then the bridge can switch from in-balance with no output potential difference to out-of-balance with an output.

18 Thermistor

This operates in a similar way to the hot wire element above, using a thermistor (see items 1 and 2, Chapter 8) instead of a wire. The change in resistance of the thermistor when it moves from above the liquid surface to below it can be much larger than that of metal wires.

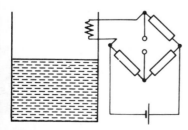

Figure 17.16 Hot wire element

18 Pressure

Pressure is defined as being the normal force per unit area. The unit of pressure is the pascal (Pa), with 1 Pa being 1 N/m². The term *absolute pressure* is used for the pressure measured relative to zero pressure, the term *gauge pressure* for the pressure measured relative to atmospheric pressure. At the surface of the earth the atmospheric pressure is generally about 100 kPa. This is sometimes referred to as a pressure of 1 bar.

Absolute pressure = gauge pressure + atmospheric pressure

Table 18.1 lists the pressure measurement systems considered in this chapter and their basic characteristics.

Further reading: Noltingk, B. E. (ed.) (1985), *Jones' Instrument Technology*, vol. 1 (*Mechanical Measurements*), Butterworth-Heinemann.

Liquid columns

Consider the gauge pressure at some depth h in a fluid at rest due to the weight of fluid above it. For a surface of area A at that depth then the force acting on it is the weight of the fluid directly above it, i.e. $hA\rho g$, where g is the acceleration due to gravity and ρ the fluid density. Thus the pressure is

$$\text{pressure } P = \frac{hA\rho g}{A} = h\rho g$$

1 U-tube manometer

The basic U-tube manometer consists of a U-tube containing a liquid. A pressure difference between the gases above the liquid in the two limbs produces a difference h in vertical heights of the liquid. For the manometer in Figure 18.1, the pressures at the bases of the two columns of liquid are $(P_1 + h_1\rho g)$ and $(P_2 + h_2\rho g)$ and since these must be equal for a fluid at rest

$$P_1 + h_1\rho g = P_2 + h_2\rho g$$

where ρ is the density of the manometric liquid and g the acceleration due to gravity. Hence

$$\text{Pressure difference} = P_1 - P_2 = h\rho g$$

If one of the limbs is open to the atmosphere then the pressure difference is that between the gas pressure and the atmosphere and is thus the gauge pressure.

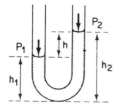

Figure 18.1   U-tube manometer

Table 18.1 Pressure measurement systems

Principle	System	Characteristics
Liquid column	1 U-tube manometer	Simple, cheap, pressures 20 Pa to 140 kPa.
	2 Industrial manometer	Can be direct reading, pressures 20 Pa to 140 kPa.
	3 Inclined tube manometer	Greater accuracy than conventional U-tube.
	4 Fortin barometer	Measures atmospheric pressure.
Diaphragms	5 Reluctance	Pressure range within 1 Pa to 100 MPa, up to 1 kHz, accuracy $\pm 0.1\%$.
	6 Capacitance	Pressure range within 1 kPa to 200 kPa, up to 1 kHz, accuracy $\pm 0.1\%$.
	7 Strain gauge	Pressure range within 0 to 100 MPa, up to 1 kHz, accuracy $\pm 0.1\%$.
	8 Force balance	Pressure range 0 to 100 kPa, accuracy $\pm 0.2\%$.
Capsules and bellows	9 Aneroid barometer	Measures atmospheric pressure.
	10 Bellows	Pressure range 200 Pa to 1 MPa, accuracy $\pm 0.1\%$, poor zero stability.
Bourdon tubes	11 Bourdon tube instruments	Pressure range within 10 kPa to 100 MPa, robust, accuracy $\pm 1\%$.
Vibrating wire	12 Diaphragm	Pressure range up to 180 kPa, accuracy $\pm 0.2\%$.
	13 Bourdon tube	Pressure range within 10 kPa to 100 MPa, robust, accuracy $\pm 1\%$.
Piezo-electric	14 Piezo-electric	Pressure range up to 500 kPa, bandwidth 5 Hz to 500 kHz.

Water, alcohol or mercury are commonly used as manometric liquids and pressure differences of the order of 20 Pa to 140 kPa can be measured. Errors can arise due to the height measured not being truly vertical, the effect of temperature on the density of the liquid, and incorrect values of the acceleration due to gravity being used. The accuracy is also affected by difficulties in obtaining an accurate reading of the level of the manometric liquid in a tube due to its meniscus. The accuracy of pressure difference measurement using a U-tube manometer is typically about $\pm 1\%$.

The correction that has to be made for the effect of temperature on the density of the manometer liquid is derived as follows. A mass m of liquid at 0 °C has a volume V_0 and a density ρ_0 related by

$$m = \rho_0 V_0$$

At temperature θ the same mass of liquid will have a volume V_θ and density ρ_θ.

$$m = \rho_\theta V_\theta$$

Hence

$$\rho_\theta V_\theta = \rho_0 V_0$$

The volume at temperature θ is related to the volume at 0 C by

$$V_\theta = V_0(1 + \gamma\theta)$$

where γ is the coefficient of cubical expansion of the liquid. Hence

$$\rho_\theta = \frac{\rho_0 V_0}{V_\theta}$$

$$\rho_\theta = \frac{\rho_0}{1 + \gamma\theta}$$

Thus, neglecting any other corrections, the pressure when measured by a manometer at temperature θ when the manometer liquid density at $0\,°C$ is known is

$$P = H\rho_\theta g = \frac{H\rho_0 g}{1 + \gamma\theta}$$

The acceleration due to gravity depends on the geographical latitude and the height above the earth's surface, decreasing by 3.086×10^{-6} m/s² for each metre above sea level. In m/s²

$$g = 9.780\,49(1 + 0.005\,288\,4 \sin^2\phi - 0.000\,005\,9 \sin^2 2\phi)$$
$$- 0.000\,003\,086H$$

where ϕ is the latitude and H the height above sea level in metres. Table 18.2 gives some typical values of g at zero height.

2 Industrial manometer

An *industrial form* of U-tube manometer (Figure 18.2) has one of the limbs with a much greater cross-sectional area than the other. A difference in pressure between the two limbs causes a difference in liquid level with liquid flowing from one limb to the other. For such an arrangement

pressure difference $= P_1 - P_2 = H\rho g$

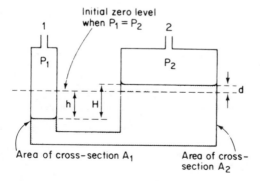

Figure 18.2 An industrial manometer

Table 18.2 Acceleration due to gravity values

Latitude	g in m/s²
0°	9.780 49
10°	9.782 04
20°	9.786 52
30°	9.793 38
40°	9.801 81
50°	9.810 79
60°	9.819 24
70°	9.826 14
80°	9.830 65
90°	9.832 21

But $H = h + d$, where h and d are the changes in level in each limb from the level that existed when there was no pressure difference. Thus

pressure difference $= (h + d)\rho g$

For the height difference the volume of liquid leaving one limb must equal the volume entering the other. Hence

$$A_1 h = A_2 d$$

where A_1 and A_2 are the cross-sectional areas of the two limbs. Hence

pressure difference $= [(A_2 d/A_1) + d]\rho g$
$= [(A_2/A_1) + 1]d\rho g$
$= [\text{a constant}]d\rho g$

Thus the movement of the liquid level d in the wide tube from its initial zero level is proportional to the pressure difference. This form of manometer thus requires only the level of liquid in one limb to be measured and this measurement is always made from the same fixed point. Usually this change in level is determined by using a float and lever system to move a pointer across a scale (see Chapter 17 for details of level measurement).

3 Inclined tube manometer

The inclined tube manometer (Figure 18.3) is a U-tube with one limb having a larger cross-section than the other and the narrower limb

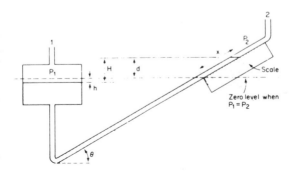

Figure 18.3 Inclined tube manometer

inclined at some angle θ to the horizontal. The vertical displacement d of the liquid level in the inclined limb is related to the movement x of the liquid along it by

$$d = x \sin \theta$$

The displacement x is measured. Thus

pressure difference $= P_1 - P_2 = [(A_2/A_1) + 1]\rho g x \sin \theta$

Since $A_2 \gg A_1$ the equation approximates to

pressure difference $= \rho g x \sin \theta$

The displacement x is greater than the vertical displacement and so gives greater accuracy than the conventional U-tube manometer.

4 Fortin barometer

The Fortin barometer (Figure 18.4) consists of a vertical glass tube in a protective jacket with scales enabling the height of the mercury surface in the tube to be directly read. First it is essential to ensure that the zero of the scale coincides with the mercury level in the lower reservoir. This mercury is in a container which has a leather bag for its lower surface and by rotating the zero adjustment screw the bag can be distorted so that the level of mercury in the reservoir coincides with the tip of the zero index. When this occurs the scale is correctly zeroed with the

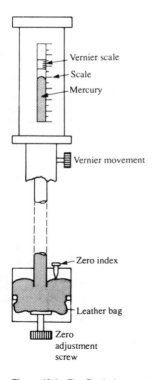

Figure 18.4 The Fortin barometer

Table 18.3 Surface tension correction

Tube diameter (mm)	Correction to be added to height (mm) for meniscus heights in mm of			
	0.2	0.5	1.0	1.5
5	0.38	0.92	1.62	
10	0.07	0.17	0.32	0.42
15	0.02	0.05	0.09	0.12
20	0.005	0.010	0.024	0.034

mercury surface. The vernier scale is then adjusted so that it just coincides with the mercury surface and so gives the height of the mercury column.

A number of corrections have to be made for an accurate value of the atmospheric pressure to be obtained, these being for the effects of temperature on the brass scale and the density of the mercury, the acceleration due to gravity value to be used, and surface tension effects. The brass scale used for the height reading will only be correct at the temperature for which it was calibrated. Readings at all other temperatures will thus be in error. If the scale has been calibrated to be correct at 0 °C and the temperature at which the reading is made is θ then

$$H_{true} = H(1 + \alpha\theta)$$

where H_{true} is the corrected reading of the scale, H is the reading at temperature θ, and α is the linear coefficient of expansion of the metal scale. The density of the mercury depends on the temperature. If γ is the coefficient of cubical expansion of the mercury then (see item 1)

$$\rho_\theta = \frac{\rho_0}{1 + \gamma\theta}$$

where ρ_θ is the density of the mercury at temperature θ and ρ_0 is the density at 0 °C. The acceleration due to gravity depends on the geographic latitude and the height above sea level (see item 1). The effect of surface tension on the mercury in the tube is to depress the level and so underestimate the reading. The amount of the depression depends on the diameter of the tube and the shape of the meniscus. Tables (see Table 18.3) are available indicating the correction that has to be applied for different diameter tubes and different meniscus heights, i.e. the height of the mercury in the centre of the tube above the height at which the mercury is in contact with the glass.

For comparison purposes it is usual to quote the height of the barometric mercury level in terms of what it would be if the measurement had been made at 0 °C and in a place where the acceleration due to gravity is 9.80665 m/s². If at temperature θ the true height, i.e. the height allowing for the scale having been calibrated at 0 °C, is h_{true}, the mercury density ρ_θ, and the acceleration due to gravity g then the pressure at that temperature is

$$\text{pressure} = h_{true}\rho_\theta g = h(1 + \alpha\theta)\rho_\theta g$$

where h is the observed height on the scale. The equivalent pressure at 0 °C is

$$\text{equivalent pressure} = h_0\rho_0 g = h(1 + \alpha\theta)\rho_\theta g$$

where h_0 is the equivalent height at 0 °C and ρ_0 the density at 0 °C.

$$\text{equivalent pressure} = \frac{h(1+\alpha\theta)\rho_0 g}{1+\gamma\theta}$$

$$= h\rho_0 g\left(1 - \frac{\{\gamma - \alpha\}\theta}{1+\gamma\theta}\right)$$

The equivalent reading of the mercury height h_0 is given by

$$h_0 = h\left(1 - \frac{\{\gamma - \alpha\}\theta}{1+\gamma\theta}\right)$$

The above is the corrected reading at a locality where the acceleration due to gravity is g, the equivalent reading where the acceleration due to gravity is $9.80665 \, \text{m/s}^2$ is

$$\text{equivalent reading} = \frac{h_0 g}{9.80665}$$

Diaphragms

With diaphragm pressure gauges a difference in pressure between the two sides of the diaphragm result in it bowing out to one side or the other. Diaphragms may be flat, corrugated or dished, the form determining the amount of displacement produced and hence the pressure range which can be measured. It also determines the degree of non-linearity. If the fluid whose pressure is required is admitted to one side of the diaphragm and the other side is open to the atmosphere then the diaphragm gauge gives the gauge pressure, if the other side is sealed then the absolute pressure is given. If fluids at different pressures are admitted to the two sides of the diaphragm then the gauge gives the pressure difference. There are a number of methods used to detect and measure the deformation.

5 Reluctance diaphragm gauge

Figure 18.5 shows the basic form of a reluctance diaphragm gauge. The displacement of the central part of the diaphragm increases the reluctance on one side of the diaphragm and decreases it on the other. The arrangement is essentially the push–pull displacement sensor described in item 6, Chapter 9 with an a.c. bridge to give an out-of-balance output related to the pressure difference causing the diaphragm displacement. The range is generally about 1 Pa to 100 MPa with an accuracy of about $\pm 0.1\%$ and a bandwidth up to 1 kHz.

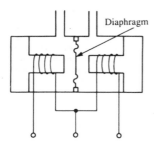

Figure 18.5 Reluctance diaphragm gauge

6 Capacitance diaphragm gauge

There are two basic forms of capacitance diaphragm gauge (Figure 18.6). With one the displacement of the diaphragm relative to a fixed plate changes the capacitance between the diaphragm and the fixed plate. The capacitor can form part of the tuning circuit of a frequency modulated oscillator and so give an electrical output related to the pressure difference across the diaphragm. With the other the diaphragm is between two fixed plates and its movement thus increases the capacitance with regard to one plate and decreases with respect to the other, i.e. a form of the push–pull displacement sensor described in item 6, Chapter 9. Such a gauge is generally used with an a.c. bridge, the out-of-balance signal being related to the pressure difference across the diaphragm. The range is generally about 1 kPA to 200 kPa with an accuracy of about ±0.1% and a bandwidth up to 1 kHz.

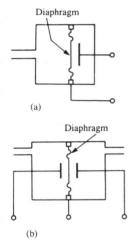

(a)

(b)

Figure 18.6 Capacitance pressure gauges

7 Strain gauge diaphragm gauge

There are a number of ways (Figure 18.7) by which strain gauges can be used to monitor the displacement of a diaphragm. One way involves them being attached to a cantilever which is bent when the central part of the diaphragm is displaced. Another way involves them being directly stuck on the diaphragm. Yet another way involves semiconductor strain gauges. While such gauges could be cemented to the surface of the diaphragm, it is now more customary to use a silicon sheet as the diaphragm and introduce doping material at appropriate places into the silicon and so produce the strain gauges integral with the diaphragm. Whatever the form of the strain gauge instrument the gauges are incorporated in a Wheatstone bridge and the out-of-balance voltage taken as a measure of the pressure difference across the diaphragm. Typically metal wire strain gauge instruments are used over the range 100 kPa to 100 MPa, with the integrated semiconductor gauge instrument used over the range 0 to about 100 kPa. The accuracy is up to about ±0.1% with a bandwidth to 1 kHz.

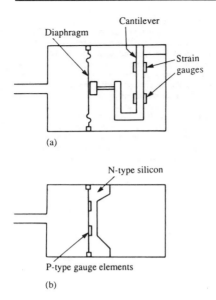

Figure 18.7 Strain gauge diaphragm gauges (a) cantilever form (b) integral semiconductor gauge form

8 Force-balance system

There are a number of versions of this form of instrument. They are all, however, based on the applied pressure causing a displacement of a diaphragm, the displacement being monitored by some transducer which produces a signal which, by means of a feedback loop, actuates a response to cancel out the displacement. A common form of such an instrument is the *pneumatic differential pressure cell* (Figure 18.8). A

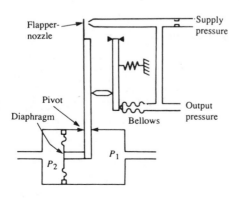

Figure 18.8 Pneumatic differential pressure cell

difference in pressure across the diaphragm results in a displacement of the diaphragm and hence the force beam. The force beam is pivoted and its movement results in a pneumatic pressure change in the flapper–nozzle system. This pressure change is communicated to a bellows, the expansion or contraction of which results in forces which act on the force beam and cancel out the diaphragm displacement. The pneumatic pressure p in the flapper–nozzle system is a measure of the pressure difference across the diaphragm,

$$p = K(P_1 - P_2) + C$$

where K and C are constants. Other forms of the cell use reluctance, LVDT or capacitive means of detecting the displacement of the diaphragm. Typically such cells have a range 0 to 100 kPa with accuracies of about $\pm 0.2\%$ and response times of about 1 s.

Capsules and bellows

9 Aneroid barometer

A capsule can be considered to be just two diaphragm. An example of an instrument employing a capsule is the aneroid barometer (Figure 18.9). It consists of a sealed capsule from which the air has been partially removed. Changes in atmospheric pressure cause the capsule to expand or contract, the displacement of the capsule surface being magnified by a system of levers to cause a pointer to move across a scale. If the capsule is completely evacuated corrections have to be made for the temperature at which the measurement is made. This is because the deflection of the capsule material and the spring depends on the elasticity of metals which in turn depends on the temperature. However if a small amount of air is left in the capsule it expands when the temperature increases and counteracts the change in the stretching properties of the metals.

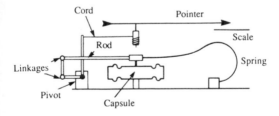

Figure 18.9 Aneroid barometer

10 Bellows pressure gauge

The length of bellows, and hence the displacement of a free end when the other end is fixed, depends on the pressure difference between the inside and outside. A number of methods are used to monitor the displacement, one form of bellows pressure sensor (Figure 18.10) uses the displacement to move one end of a pivoted lever, another moves the core of an LVDT, while another uses the movement of the end of the bellows to move the slider of a potentiometer. Bellows instruments are used to measure pressure differences in the range 200 Pa to 1 MPa with an accuracy of about $\pm 0.1\%$. They have poor zero stability.

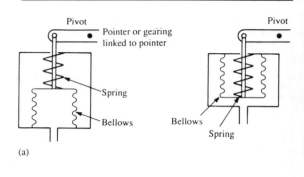

(a)

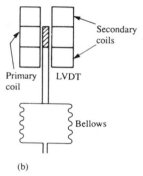

(b)

Figure 18.10 Bellows pressure gauge (a) lever types (b) LVDT type

Bourdon tubes

The Bourdon tube (see item 24, Chapter 8) may be in the form of a 'C', a flat spiral, a helical spiral, or twisted; an increase in the pressure inside the tube causing the tube to straighten out to an extent which depends on the pressure change.

11 Bourdon tube instruments

The displacement of the end of the tube may be monitored in a variety of ways, e.g. to directly move a pointer across a scale, via gearing to move a pointer across a scale, to move the slider of a potentiometer or to move the core in a LVDT (Figure 18.11). Bourdon tube instruments typically operate in the range 10 kPa to 100 MPa, the range depending on whether the tube is C, helical, or twisted, and on the material from which it is made. C-shaped tubes made from brass or phosphor bronze have a pressure range from about 35 kPa to 100 MPa. Spiral and helical tubes are more expensive and have greater sensitivity but as a consequence a lower maximum pressure that can be measured, typically about 50 MPa. Bourdon tubes are robust with an accuracy of about ±1% of full scale reading. The main sources of error are hysteresis, changes in sensitivity due to temperature changes, and frictional effects with linkages and pointers.

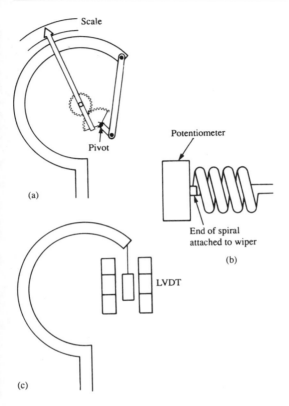

Figure 18.11 Bourdon tube instruments (a) geared form (b) potentiometer form (c) LVDT form

Vibrating wire pressure gauges

The free vibration frequency f of a stretched wire is given by

$$f = \frac{1}{2L}\left(\frac{T}{m}\right)^{\frac{1}{2}}$$

where T is the wire tension, L its vibrating length and m the mass per unit length. The wire is set into vibration by an actuator fed by an oscillator. The frequency of the oscillator is adjusted until the amplitude of the vibrating wire is a maximum, the amplitude being monitored by some sensor, e.g. a variable reluctance pick-up. The frequency is thus a measure of the tension in a wire.

12 Diaphragm gauge

A wire is stretched between a fixed support and the centre of a diaphragm (Figure 18.12). Movement of the diaphragm as a result of pressure changes causes the tension in the wire to change. Pressures up to 180 kpa are typically measured by this method, the accuracy being as high as $\pm 0.2\%$.

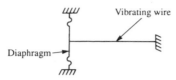

Figure 18.12 Vibrating wire diaphragm gauge

13 Bourdon gauge

The movement of the end of a C-type Bourdon gauge can be used to alter the tension in a stretched wire and hence the frequency with which it freely vibrates. Another form has a helical Bourdon gauge: the rotation of the end of the Bourdon tube as a result of a pressure change is used to alter the tension in a stretched wire (Figure 18.13). The range and accuracy of such gauges is typically that of Bourdon gauges.

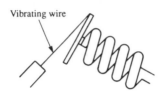

Figure 18.13 Vibrating wire Bourdon gauge

Piezo-electric pressure gauges

See item 15, Chapter 8 for a discussion of piezo-electricity.

14 Piezo-electric electrostatic gauge

A piezo-electric pressure gauge consists essentially of a diaphragm which presses against a piezo-electric crystal. Movement of the diaphragm causes the crystal to be compressed and a potential difference is produced across its faces. Typically, such a gauge can be used for pressures from about 200 kPa to 100 MPa and a bandwidth of 5 Hz to 500 kHz. It cannot be used for static pressures.

Calibration

Calibration can be by means of a manometer or *dead-weight tester* (Figure 18.14). With the latter the pressure is produced in a fluid by winding in a piston. The pressure is determined by means of adding weights to the platform so that it remains at a constant height. If the total mass of the platform and its weights is M then its weight is Mg. If the cross-sectional area of the platform piston is A then the pressure is Mg/A.

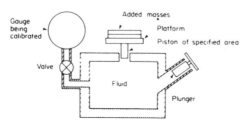

Figure 18.14 Dead-weight tester

19 Radiation

Radiation detectors are used to detect and measure the radiations emitted by radioactive materials and produced by nuclear accelerators, these radiations including alpha particles, beta particles, gamma rays and a variety of other particles. Table 19.1 lists the instruments discussed in this chapter and their main characteristics.

Further reading: Noltingk, B. E. (ed.) (1987), *Jones' Instrument Technology*, vol. 3, (*Electrical and Radiation Measurements*), Butterworth-Heinemann.

Gas detectors

1 Ionization chamber

An ionization chamber is a gas (often air) filled chamber containing two electrodes, generally a cylindrical cathode with a central wire anode, or a pair of parallel plates. When the radiation enters the chamber it ionizes the gas. When the potential difference between the electrodes is small many of the ions recombine before they reach the electrodes. If the potential difference is increased less recombination occurs until a voltage is reached when all the ions reach the electrodes without recombining. This is the voltage at which the chamber is operated, the ionization current between the electrodes then being of the order of 10^{-15} to 10^{-10} A. This current is a measure of the rate of production of ion pairs in the gas in the chamber. Accuracy of the order of $\pm 0.1\%$ is possible and the output is little affected by the energy of the radiation.

A particular use of ionization chambers is as personnel dose meters. The two electrodes form a capacitor which is charged by a potential difference to the required voltage. The electrodes are then isolated from the supply. The ionization current resulting from radiation passing through the chamber discharges the capacitor, the loss of charge of the capacitor over a period of time then becoming a measure of the total number of ions pairs produced in that time. Figure 19.1 shows the form of the *pocket dosimeter*. This has a central wire

Table 19.1 Radiation detectors

Principle	System	Characteristics
Gas detectors	1 Ionization chamber	Gives a current reading, high accuracy, can be used for dose measurement.
	2 Geiger counter	Detects individual ionizing particles.
Scintillation	3 Scintillation counter	Detects individual particles, type of radiation determines scintillator used.
Photographic film	4 Film badge	Measures dose.
Thermo-luminescence	5 Thermo-luminescent dosimeter	Measures dose, can be used many times.

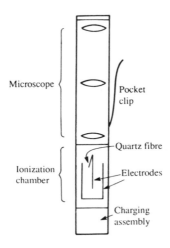

Figure 19.1 Pocket dosimeter

electrode surrounded by a cylindrical electrode. The electrodes are charged to the required voltage when the dosimeter is plugged into an external charging unit. The state of charge of the central electrode is indicated by a quartz fibre which is attached at one end to it. The greater the charge on the electrode the more the fibre deflects, the fibre being viewed against a scale.

2 Geiger counter

If the voltage between a pair of electrodes in a gas is increased beyond the point at which all the ions are being collected, the electrons begin to be accelerated sufficiently by the voltage to have enough energy to ionize further gas atoms by collision. With sufficient voltage the electrons produced in this secondary ionization can cause further ionization by collision. This 'avalanche' process can produce multiplication of the original ionization by factors of the order of 10^3 to 10^5. When the multiplication is of the order of 10^3 to 10^4 the current pulse produced by a single ionization event is proportional to the energy lost by the particle responsible for the initial ionization. A counter operating in this region is called a *proportional counter*. At higher multiplications the output current pulse becomes independent of the energy of the particle causing the initial ionization. This is the region used for the *Geiger counter*. These instruments are referred to as counters because the pulses due to individual ionizing particles can be registered. In operating a Geiger counter the voltage between the electrodes is increased until the indicated count rate reaches a plateau (Figure 19.2) and becomes virtually independent of the voltage. The counter is then operated at a voltage in roughly the centre of the plateau so that any fluctuations in the voltage will have little effect on the indicated count rate.

Figure 19.3 shows the basic form of a Geiger tube. It is a tubular envelope of glass or metal with a central wire anode and an outer cylindrical cathode, either a metal cylinder or a coating of conducting material on the inside of the envelope. The window of the Geiger tube

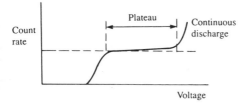

Figure 19.2 Geiger counter characteristic

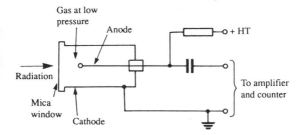

Figure 19.3 Geiger counter

has to be selected for the energy and type of radiation involved. Thus for alpha particles and low energy beta particles a very thin window is required. The gas in the tube is usually argon at a pressure of about 10 kPa, with the addition of chlorine or bromine vapour to act as a quenching agent. These atoms dissociate when hit by electrons and so mop up energy and limit the time for which the avalanche effect lasts. The output pulse is of the order of 1 V. There is a time between the initiation of an avalanche in the tube by an ionizing particle and when another ionizing particle can trigger another avalanche of charge. This is called the *dead time* and is of the order of 100 μs. Corrections have thus to be made to the count rate R indicated by a Geiger counter for the counts lost during this dead time τ.

$$\text{True count rate} = \frac{R}{1 - R\tau}$$

Scintillation counter

3 Scintillation counter

The basic scintillation counter consists of a scintillator, a photomultiplier and an amplifier (Figure 19.4). When an ionizing particle passes into the scintillator it produces a flash of light. This is detected and amplified by a photomultiplier, before further amplification by an electronic amplifier to give a pulse which can be counted. A photomultiplier (see item 18, Chapter 8 and Figure 8.20(b)) is a tube where the light falls on a light sensitive cathode which then emits electrons. These electrons are then accelerated by a potential difference to another electrode where secondary emission results in yet more electrons. Perhaps ten such accelerations will occur with multiplication of the electrons at each. The scintillator may be a single

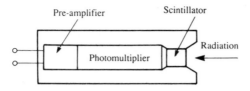

Figure 19.4 Scintillation counter

inorganic crystal, e.g. sodium iodide with thallium iodide additive (referred to as NaI(Tl)), a single organic crystal, e.g. anthracene, a plastic loaded with small amounts of other materials, or a liquid solution in which the radioactive sample is directly mixed with the liquid scintillator. The choice of scintillator is determined by the type and energy of the radiations to be detected. Another factor that has to be considered is the time for the light to decay, i.e. the scintillation duration, since this determines whether it can respond to high count rates. The NaI(Tl) crystal is widely used, being good for the detection of gamma radiation and X-rays, giving a light output which is proportional to the energy of the radiation. The CsI(Tl) crystal is widely used for alpha particles. Scintillation counters, with suitable choice of scintillator, can be used for measurement of alpha, beta, gamma, X-ray and neutron radiations. It is particularly useful for gamma and X-ray radiation where it has an efficiency approaching 100%.

Photographic film

Ionizing radiations cause, after development, blackening of photographic emulsions. Such emulsions can be used to detect and show the path of ionizing radiations since the blackening occurs where the ionization occurs.

4 Film badge

Figure 19.5 shows the form of the film badge dosimeter. Photographic emulsions of different sensitivities coat opposite sides of the cellulose–acetate base. To enable the energy and type of radiation to be

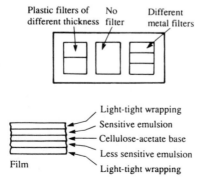

Figure 19.5 Film badge

estimated, the film is partially covered with filters of different materials and thicknesses so that the degree of blackening under the filters can be compared with that where no filter was used. For example, large differences in blackening suggest low-energy radiation while small differences high energy radiation.

Thermoluminescence

When radiation passes through certain materials, such as lithium fluoride and calcium fluoride, some of the energy becomes stored in its atoms. This radiation is released as visible light when the material is subsequently heated. This effect is known as *thermoluminescence*.

5 Thermoluminescence dosimeter

The dosimeter consists of a material such as lithium fluoride bonded to a strip of plastic. After exposure the dosimeter is sandwiched between a heat source and a photomultiplier. On heating, the count rate due to the thermoluminescence is recorded. The dosimeter can be re-used many times.

20 Stress and strain

Tensile and compressive *stress* is defined as the force acting per unit cross-sectional area of a material and which results in an increase or decrease in length. *Strain* is defined as the change in length per unit length. The unit of stress is the pascal (Pa) and strain being a ratio has no units. Measurements are generally made of the strain and the stress deduced from them. Table 19.1 lists the methods for strain measurement discussed in this chapter and their main characteristics.

Further reading: Noltingk, B. E. (ed.) (1985), *Jones' Instrument Technology*, vol. 1 (*Mechanical Measurements*), Butterworth-Heinemann; Holister, G. S. (1967), *Experimental Stress Analysis, Principles and Methods*, Cambridge University Press.

Extensometers

Extensometers are used to directly measure the change in length of a gauge length and are thus in fact measuring very small displacements.

1 Huggenberger extensometer

The Huggenberger extensometer (Figure 20.1) uses a compound lever to give a high magnification, often 2000 or more. Extension of the specimen between the gauge points results in rotation of the first lever about its pivot point. Since the input to this first lever is very close to the pivot point and the output a comparatively large distance away at the far end of the lever, a large magnification of the input movement is transmitted to the second lever, at a point close to its pivot. The output from that lever is the movement of the pointer across the scale.

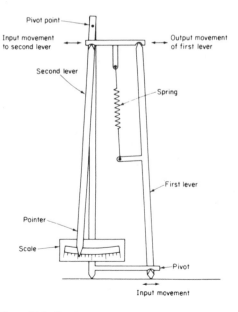

Figure 20.1 The Huggenberger extensometer

Table 20.1 Measurement of strain

Principle	System	Characteristics
Extensometer	1 Huggenberger	Uses compound lever for high magnification.
	2 Johansson	Uses twisted metal strip for high magnification.
Strain gauge	3 Resistance	Strain measured over small gauge length, temperature compensation required.
	4 Vibration	Large gauge length, high stability, used for internal stresses in concrete.
Whole surface	5 Brittle lacquer	Used for determining the directions and locations of maximum stresses.
	6 Moiré fringes	Often used for large strains.
Photoelasticity	7 Photoelasticity	Used for models made with certain materials.

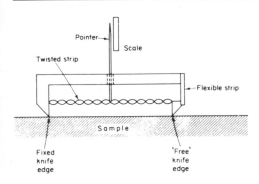

Figure 20.2 Johansson extensometer

2 Johansson extensometer

The basis of this extensometer (Figure 20.2) is a twisted metal strip. When the material extends, the free knife edge rotates and changes the degree of twist in the metal strip. As a result a pointer moves across the scale. The twisted strip is able to give magnifications of the order of 5000.

Strain gauges

3 Electrical resistance strain gauge

The electrical resistance strain gauge (see item 3, Chapter 8 and Figure 8.3) is stuck to the surface of the test material so that when the surface is subject to strain the gauge suffers the same strain. The strain results in a change in the resistance (δR) of the strain gauge.

$$\frac{\delta R}{R} = G \times \text{strain}$$

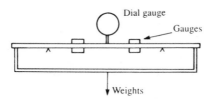

Figure 20.3 Strain gauge calibration

where R is the original resistance of the gauge and G a constant for the gauge called the gauge factor. For most metal wire or foil strain gauges the gauge factor has a value of about 2. The gauge factor is supplied by the strain gauge manufacturer from a calibration made of a number of gauges taken from the same production batch. Such a calibration can involve the strain gauges being attached to a tensile test piece and the strain determined using an extensometer. Another method is to use four point bending of a bar to which the gauges have been stuck (Figure 20.3). With this method the radius of curvature R of the bar is determined from a measurement of the deflection of the bar midpoint and the distance y of the strain gauge element from the neutral axis of the bar. It is important to take into account the thickness of the gauge backing and the adhesive layer when making this measurement.

$$\text{Strain} = \frac{y}{R}$$

Changes in temperature can produce similar changes in resistance to those resulting from the strain. The effects of temperature can be eliminated by using a dummy gauge. This is a strain gauge of the same resistance as the gauge being strained and mounted on a piece of the same material. This however is not subject to the strain, merely positioned close to the active gauge so that it is subject to the same temperature changes. The active and dummy gauges are connected in adjacent arms of a Wheatstone bridge (see item 3, Chapter 9 and Figure 9.5). Thus the effects of temperature changes on the two gauges cancel out, leaving only the difference between the two resulting from the strain on the active gauge.

Self-temperature compensated strain gauges are available for certain materials. The gauge material is chosen so that its resistivity change with temperature cancels out the effect of the thermal expansion of the material to which the gauge is attached.

Strain gauges measure the strain in the direction of the length of the gauge wires or foil elements. If there is a uniaxial stress and the strain gauge is aligned along this axis then

$$\text{stress} = E \times \text{strain}$$

where E is the tensile modulus of the material to which the gauge is attached. However, a strain gauge on the surface at right angles to this uniaxial stress would indicate a strain despite there being no stress applied in this direction, this being

$$\text{transverse strain} = -v \times \text{longitudinal strain}$$

where v is Poisson's ratio.

If the stress at the surface is biaxial the principal stresses are mutually perpendicular to each other. Two strain gauges are thus needed, these being at right angles to each other and aligned in the principal stress directions

$$\text{stress in } x \text{ direction} = \frac{E(\varepsilon_x + v\varepsilon_y)}{1 - v^2}$$

$$\text{stress in } y \text{ direction} = \frac{E(\varepsilon_y + v\varepsilon_x)}{1 - v^2}$$

where ε_x is the strain in the x direction and ε_y in the y direction, E the tensile modulus and v Poisson's ratio. If the directions of the principal stresses are not known it is necessary to use three strain gauges in what is termed a rosette (Figure 20.4). If the gauges are at 45° and 90° to each other than the principal strain ε_x and ε_y are related to the strains ε_1, ε_2 and ε_3 indicated by the three gauges

$$\varepsilon_1 = \left(\frac{\varepsilon_x + \varepsilon_y}{2}\right) + \left(\frac{\varepsilon_x - \varepsilon_y}{2}\right)\cos 2\theta$$

with θ being the angle between the ε_1 and ε_y directions.

$$\varepsilon_2 = \left(\frac{\varepsilon_x + \varepsilon_y}{2}\right) + \left(\frac{\varepsilon_x - \varepsilon_y}{2}\right)\cos 2(\theta + 45^\circ)$$

$$\varepsilon_3 = \left(\frac{\varepsilon_x + \varepsilon_y}{2}\right) + \left(\frac{\varepsilon_x - \varepsilon_y}{2}\right)\cos 2(\theta + 90^\circ)$$

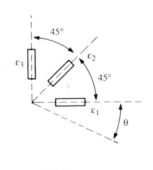

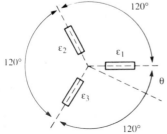

Figure 20.4 Strain gauge rosettes

If the gauges are at 120° to each other,

$$\varepsilon_1 = \left(\frac{\varepsilon_x + \varepsilon_y}{2}\right) + \left(\frac{\varepsilon_x - \varepsilon_y}{2}\right)\cos 2\theta$$

$$\varepsilon_2 = \left(\frac{\varepsilon_x + \varepsilon_y}{2}\right) + \left(\frac{\varepsilon_x - \varepsilon_y}{2}\right)\cos 2(\theta + 120°)$$

$$\varepsilon_3 = \left(\frac{\varepsilon_x + \varepsilon_y}{2}\right) + \left(\frac{\varepsilon_x - \varepsilon_y}{2}\right)\cos 2(\theta + 240°)$$

The principal stresses can then be calculated using these principal strains and the equations given earlier.

Further reading: Holister, G. S. (1983), *Experimental Stress Analysis*, Cambridge Engineering Services; Window, A. L. and Holister, G. S. (1982), *Strain Gauge Technology*, Allied Science Publishers.

4 Vibrating wire strain gauge

A wire stretched between two supports will freely vibrate with a fundamental frequency f given by

$$f = \frac{1}{2L}\sqrt{\frac{T}{m}}$$

where L is the length of the vibrating wire, T its tension and m its mass per unit length. As the wire obeys Hooke's law then the tension is given by

$$T = k\delta L$$

where δL is the amount by which the wire has been stretched and k a constant. If the ends of the wire are bonded to the material being strained then both L and T will change and so the frequency f can be used as a measure of the strain. The amplitude of the vibrating wire is monitored by means of a pick-up coil, the wire being ferromagnetic, and the frequency then determined from the amplitude variation with time. Such a gauge (Figure 20.5) tends to be rather large, of the order of 100 mm, and has good stability. It is used for the measurement of internal strains in concrete.

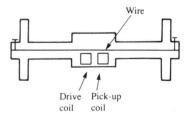

Figure 20.5 Vibrating wire strain gauge

Whole surface strain measurement

5 Brittle lacquers

The surface of the test piece is coated with a special lacquer. When the lacquer dries it becomes brittle and when the test piece is stressed cracks appear where the strain is above a certain threshold. The cracks form at right angles to the maximum principal stresses and thus the pattern of cracks show the location of the highest stresses. The method is of most use for determining the directions of the principal stresses, the accuracy in relation to their magnitude is very low.

6 Moiré fringes

A line pattern is etched on the surface of the test piece. A transparent grating with the same number of lines per unit length is then held over the surface and the surface line pattern viewed through it. A Moiré fringe pattern is seen, the pattern changing as the test piece is subject to strain and the spacing of its line pattern changes.

Photoelasticity

A beam of light can be considered to travel as a wave which can have planes of vibration at any number of directions at right angles to the axis of propagation, i.e. it is a transverse wave. Normally light consists of waves with many different planes of vibration. If the wave is however confined to just one plane of vibration it is said to be plane polarized.

7 Photoelasticity

Photoelasticity is a technique for determining stresses in models made of certain materials, e.g. Perspex. Such materials have the property called *birefringence*. If plane polarized light is passed through a birefringent material, then when that material is stressed the ray of light splits into two plane polarized components. These have their planes of vibration at right angles to each other, the planes being along the directions of the principal stresses. The velocity of each of the components is proportional to the size of the stress in its plane of vibration. Thus in passing through the stressed material the waves get out of step if the stresses differ. When the waves emerge from the material and are recombined (Figure 20.6), interference fringes are produced.

When the two principal stresses are both zero the waves do not get out of step. If the differences in stresses is such that the waves get out of step by just enough to cancel each other out, then an interference fringe is produced. This first out of step fringe is said to have the order number 1. The stresses may however be sufficient for the waves to get out of step by more than this, higher order fringes then being produced.

Principal stress difference = nf/t

where n is the fringe order number, t the thickness of the model and f a constant for the model material called the material fringe value. Thus by determining the order number of the fringe at a point the difference in principal stress values can be obtained. One way of doing this is to count the number of fringes occurring between the point concerned and a zero stress position for the model.

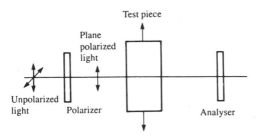

Figure 20.6 Photoelasticity

21 Temperature

Table 21.1 outlines the temperature measurement systems discussed in this chapter and their characteristics.

Further reading: Noltingk, B. E. (ed.) (1985), *Jones' Instrument Technology*, vol. 2 (*Measurement of Temperature and Chemical Composition*), Butterworth-Heinemann.

Table 21.1   Temperature measurement systems

Principle	System	Characteristics
Expansion	1 Bimetallic strip	Can be used for thermostats, range within -30 to $+600\,°C$, accuracy $\pm 1\%$, robust.
	2 Liquid-in-glass	Direct reading, range depends on liquid, with mercury -35 to $+600\,°C$, alcohol -80 to $+70\,°C$, pentane -200 to $+30\,°C$, general accuracy $\pm 1\%$, fragile.
	3 Liquid-in-metal	Range depends on liquid, within $-90\,°C$ to $+650\,°C$, e.g. mercury $-39\,°C$ to $+650\,°C$, accuracy $\pm 1\%$, robust, distant reading.
	4 Gas-in-metal	Range $-100\,°C$ to $+650\,°C$, accuracy $\pm 0.5\%$, robust, distant reading.
	5 Vapour pressure	Range within $0\,°C$ to $250\,°C$, accuracy $+1\%$, robust, distant reading.
Resistance	6 Metal	Range depends on metal, for platinum $-200\,°C$ to $+850\,°C$, nickel $-80\,°C$ to $+300\,°C$, copper $-200\,°C$ to $+250\,°C$, accuracy for platinum $\pm 0.5\%$.
	7 Thermistor	Non-linear, range within $-100\,°C$ to $+300\,°C$, fast response.
Thermo-electric	8 Thermo-couple	Range, sensitivity and accuracy depends on metals used, e.g. iron–constantan $-180\,°C$ to $760\,°C$, $53\,\mu V/°C$, ± 1 to 3%; platinum–platinum/rhodium 13% $0\,°C$ to $1750\,°C$, $6\ /\ \mu V/°C$, $< \pm 1\%$
Pyrometer	9 Disappearing filament	Range $+600\,°C$ to $3000\,°C$, accuracy $\pm 0.5\%$, no physical contact with hot object.

continued

Table 15.1 (*continued*)

Principle	System	Characteristics
	10 Radiation	Range 0 °C to 3000 °C, accuracy ±0.5%, no physical contact with hot object.
	11 Two colour	Range 0 °C to 3000 °C, accuracy ±0.5%, no physical contact with hot object.

The *International Practical Temperature Scale* uses a number of fixed points, for which numerical values are given, and specifies what thermometers are to be used to determine the temperatures in the intervals between the fixed points (Table 21.2). Temperatures on this scale can be expressed in two different ways, in Celsius degrees (°C) or kelvins (K).

Temperature in K = temperature in °C + 273.15

Table 21.2 The International Practical Temperature Scale fixed points

Fixed point	Temperature °C	K	Interpolation thermometer
Triple point of hydrogen	− 259.34	13.81	
Boiling point of hydrogen at 33 330.6 Pa pressure	− 255.478	17.042	Platinum
Boiling point of hydrogen	− 252.24	20.28	resistance thermometer
Boiling point of neon	− 246.048	27.102	
Triple point of oxygen	− 218.789	54.361	
Triple point of argon	− 193.352	83.798	
Boiling point of oxygen	− 186.962	90.188	
Triple point of water	0.01	273.16	
Boiling point of water	100	373.15	Platinum
Freezing point of tin	231.9681	505.1181	res. therm.
Freezing point of zinc	419.58	692.73	
	630.74	903.89	
			Thermocouple
Freezing point of gold	1064.43	1337.58 upwards	Radiation pyrometer

Note: The above are termed the primary fixed points, there are also secondary fixed points to aid in specifying temperatures within the intervals between the primary fixed points. Unless otherwise stated the boiling points are at normal atmospheric pressure, i.e. 101 325 Pa. The triple point is the temperature at which the solid, liquid and gas coexist. Below 0 °C the resistance–temperature relationship for the platinum resistance thermometer is established using a reference

function and specified deviation equations. Above 0 °C two poly-
nomial equations are used for the platinum resistance thermometer. A
quadratic equation is specified for the thermocouple, platinum–10%
rhodium/platinum, and for the radiation thermometer the Planck law
of radiation is used.

Expansion types

1 Bimetallic strips

The bimetallic strip consists of two different metal strips of the same
length bonded together. Because the metals have different coefficients
of expansion a temperature change results in the curvature of the strip
changing, the metal with the larger coefficient of expansion being on
the outside of the curve. The amount by which the strip curves depends
on the two metals used, the length of the composite strip, and the
change in temperature. If one end of a bimetallic strip is fixed the
amount by which the other end moves is a measure of the temperature.
This movement may be used to open or close electrical circuits, as in
the simple thermostat used with many domestic central heating
systems (Figure 21.1). Because the longer the length of the bimetallic
strip the greater the movement, bimetallic strip thermometers usually
have the strip wound in the form of a helix (Figure 21.2). Movement of
the free end is then used to directly move a pointer across a scale.
Bimetallic strip devices are robust, relatively cheap, can be used within
a range of about −30 °C to 600 °C, can be used for thermostats, have
an accuracy of the order of ±1%, but are fairly slow reacting to
change, and are relatively cheap.

2 Liquid in glass thermometers

The liquid in glass thermometer involves a liquid expanding up a
capillary tube. Such thermometers are direct reading, fragile, capable
of reasonable accuracy under standardized conditions, fairly slow
reacting to change and cheap. With mercury as the liquid the range is
−35 °C to +600 °C, with alcohol −80 °C to +70 °C, with toluene
−80 °C to +100 °C, with pentane −200 °C to +30 °C, with creosote
−5 °C to +200 °C. Thermometers are calibrated for use partially

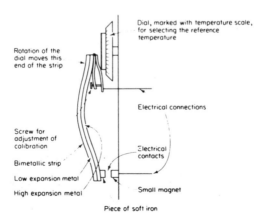

Figure 21.1 Bimetallic thermostat

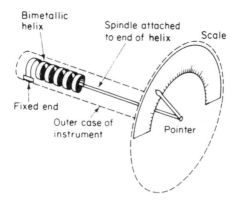

Figure 21.2 Bimetallic thermometer

immersed up to some particular mark on the stem, totally immersed when the thermometer is immersed to the level of the liquid in the thermometer stem, and completely immersed when the entire thermometer is immersed. If a thermometer is not immersed to the amount for which it was calibrated then errors occur. In general, accuracy is about $\pm 1\%$.

3 Liquid in metal thermometers

Liquid in metal thermometers consist of a metal bulb containing a liquid connected to a Bourdon tube by a capillary tube. When the liquid expands there is an increase in pressure which is then registered by a Bourdon tube pressure gauge. With mercury the range is $-39\,°C$ to $+650\,°C$, with alcohol $-46\,°C$ to $+150\,°C$, xylene $-40\,°C$ to $+400\,°C$, ether $+20\,°C$ to $+90\,°C$. A variety of other liquids are also used, in general the ranges fall within $-90\,°C$ to $+650\,°C$. The instrument can be used to give readings at a distance from the thermometer bulb. Accuracy is about $\pm 1\%$.

A source of error with this type of thermometer is the liquid in the connecting capillary tube, the temperature of this having an effect on the pressure. The error is reduced by making the volume of this small, hence the use of capillary tubing. Another way is to have a second capillary tube alongside the main capillary tube but terminating just before the bulb (Figure 21.3). It is connected to a second Bourdon tube and the display pointer is driven by the difference in movement between the two Bourdon tubes. Another method involves the use of a bimetallic strip. The strip is connected to the end of the Bourdon tube and causes it to be displaced by an amount which depends on the ambient temperature and compensates for the liquid in the capillary tube.

Other sources of error are head errors, ambient pressure errors and immersion errors. Head errors occur if the height of the thermometer bulb changes with respect to the Bourdon tube. This is due to the height of the liquid in the thermometer exerting a pressure. The Bourdon gauge measures the gauge pressure and thus changes in the ambient pressure will affect its readings. The thermometer bulb needs to be fully immersed if correct readings are to be obtained.

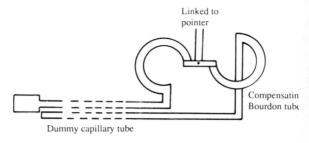

Figure 21.3 Ambient temperature compensation

4   Gas in metal thermometer

The industrial form of a gas thermometer consists of a thermometer bulb connected to a Bourdon gauge and filled with gas, e.g. nitrogen. When the temperature rises the gas pressure increases and is indicated by the Bourdon gauge. The bulb of the thermometer is fairly large, about 50 to 100 cm^3. The thermometer is robust, has a range of about $-100\,°C$ to $650\,°C$, is direct reading, can be used to give a display at a distance, and has an accuracy of about $\pm 0.5\%$ of full scale deflection. The sources of error are the same as those described in item 3 above, the head error being, however, negligible.

5   Vapour pressure thermometers

The vapour pressure thermometer consists of a thermometer bulb connected to a Bourdon gauge and partially filled with liquid, the space above the liquid containing only the vapour from the liquid. The higher the temperature the greater the amount of liquid that has evaporated and the greater the pressure exerted by its vapour. The vapour pressure is indicated by the Bourdon gauge and is a measure of the temperature. The pressure–temperature relationship is however not linear. With methyl chloride as the liquid the range is about $0\,°C$ to $50\,°C$, with sulphur dioxide $30\,°C$ to $120\,°C$, di-ethyl ether $60\,°C$ to $160\,°C$, ethyl alcohol $30\,°C$ to $180\,°C$, water $120\,°C$ to $220\,°C$, toluene $150\,°C$ to $250\,°C$. The instrument is robust, direct reading, can be used at a distance (greater than that possible with liquid-in-metal thermometers), has a non-linear scale and an accuracy of about $\pm 1\%$. The sources of error are the same as those described in item 3 above.

Resistance thermometers

6   Metal resistance thermometers

The resistance of metals generally increases with temperature, the change in resistance being proportional to the temperature change (see item 1, Chapter 8). The resistance thermometer consists of a coil of wire connected to a circuit, usually a Wheatstone bridge, for the monitoring of its change in resistance. The coil generally consists of the resistance wire wound over a ceramic coated tube, it then being also coated with ceramic, and mounted in a protecting tube. The response time is fairly slow, often of the order of a few seconds, because of the poor thermal contact between the coil and the medium whose temperature is being measured. The metals mainly used for the resistance coil are platinum, nickel and copper (see Figure 8.1).

Platinum has a closely linear resistance–temperature relationship, gives good repeatability, has long term stability, can give an accuracy of $\pm 0.5\%$ or better, has a temperature range of about $-200\,^{\circ}\mathrm{C}$ to $+850\,^{\circ}\mathrm{C}$, is relatively inert and so can be used in a wide range of environments without deterioration, but is more expensive than many other metals. It is however the most widely used metal. The temperature coefficient of resistance α is about $0.0039/^{\circ}\mathrm{C}$. Nickel and copper are cheaper but have less stability, are more prone to interaction with the environment and cannot be used over such a large range of temperature. Nickel has a temperature coefficient of resistance α of about $0.0067/^{\circ}\mathrm{C}$ and a range of about -80 to $+300\,^{\circ}\mathrm{C}$. Copper has a temperature coefficient of resistance α of $0.0038/^{\circ}\mathrm{C}$ and a range of about -200 to $+250\,^{\circ}\mathrm{C}$.

A problem with using a Wheatstone bridge is that the resistance measured will be that of the resistance coil and the leads connecting it to the bridge. If the temperature of the leads change then there will be a resistance change, regardless of what happens to the temperature of the resistance element. Methods for compensating for this are discussed in item 3, Chapter 9, see Figures 9.3 and 9.4.

7 Thermistor

Thermistors give much larger resistance changes per degree than metal wire elements, however the resistance–temperature relationship is very non-linear (see item 1, Chapter 8). Their small size, often just a small bead, means a small thermal capacity and hence a rapid response to temperature changes. The temperature range over which they can be used will depend on the thermistor concerned, ranges within about $-100\,^{\circ}\mathrm{C}$ to $+300\,^{\circ}\mathrm{C}$ are possible. Over a small range the accuracy can be $0.1\,^{\circ}\mathrm{C}$ or better, however their characteristics tend to drift with time. A Wheatstone bridge can be used for the resistance measurement, there is however no need for compensation for lead resistance since the resistance of the leads is negligible compared with that of the thermistor.

Thermoelectric effect

See item 14, Chapter 8 for a discussion of this effect.

8 Thermocouples

See item 14, Chapter 8 for details and data regarding thermocouples. Thermocouples have very small thermal capacity and so respond rapidly to changes in temperature. The base metal thermocouples, E, J, K and T, are relatively cheap, with accuracies of about ± 1 to 3%, but deteriorate with age. The noble metal thermocouples, R and S, are more expensive, with accuracies of the order of $\pm 1\%$ or better, and are more stable with long life. Standard tables are available which give the e.m.f.s of commonly used thermocouples as a function of temperature when one junction is at $0\,^{\circ}\mathrm{C}$. Item 14, Chapter 8 includes extracts from such tables. The reference junction of the thermocouple is usually at $0\,^{\circ}\mathrm{C}$, e.g. by immersing it in a mixture of ice and water. An alternative to this is to include in series with the thermocouple a circuit which gives a potential difference which just compensates for the junction being at some other temperature than $0\,^{\circ}\mathrm{C}$. See item 14, Chapter 8 for details of such a circuit. The e.m.f. of a thermocouple can be measured by directly connecting it to a galvanometer, or a potentiometer circuit (as in item 8, Chapter 9), or an electronic circuit involving a high impedance amplifier.

A larger output per degree is given by using a group of thermocouples connected in series so that the e.m.f.s from each add, such an arrangement being known as a *thermopile*.

Pyrometers

Figure 21.4 shows how the power P_λ emitted per unit surface area at a particular wavelength λ from a black body varies across the wavelengths at different temperatures T (on kelvin scale). The distribution is described by *Planck's law*.

$$P_\lambda = \frac{c_1}{\lambda^5[\exp(c_2/\lambda T) - 1]}$$

where c_1 and c_2 are constants. The total power P per unit surface area emitted at a particular temperature is the area under the appropriate graph, i.e. the integral of the above expression from zero to infinity. This is given by

$$P = \sigma T^4$$

where σ is a constant, called the Stefan–Boltzmann constant.

The above all refers to a black body. Such a body is a theoretical ideal. A real body emits less radiation at any particular wavelength or temperature and a correction factor called the emissivity ε is necessary.

$$\varepsilon = \frac{\text{actual radiation at } \lambda \text{ and } T}{\text{black body radiation at } \lambda \text{ and } T}$$

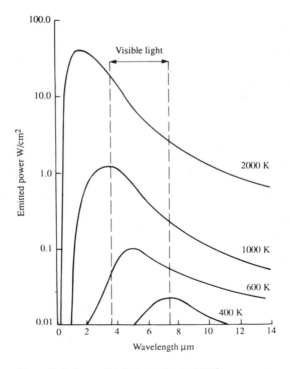

Figure 21.4 Power distribution with wavelength

Table 21.3 Emissivity values

Material	Condition	Temperature °C	Emissivity
Aluminium	oxidized	600	0.2
Brass	oxidized	600	0.6
Cast iron	oxidized	600	0.8
Cast iron	strongly oxidized	250	0.95
Copper	oxidized	200	0.6
Fireclay		1000	0.6

The emissivity of a material depends on the surface shape and texture, its temperature, and the wavelength. Table 21.3 shows some approximate values.

9 Disappearing filament pyrometer

The disappearing filament pyrometer (Figure 21.5) involves just the visible part of the radiation emitted by a hot object. The radiation is focused onto a filament so that the radiation and the filament can both be viewed in focus through an eyepiece. The filament is heated by an electrical current until the filament and the hot object seem to be the same colour, i.e. the filament disappears into the background of the hot object. The filament current is then a measure of the temperature. A red filter between the eyepiece and the filament is generally used to make the matching of the colours of the filament and the hot object easier. Another red filter may be introduced between the hot object and the filament with the effect of making the object seem less hot in comparison with the filament and so extending the range of the instrument. The disappearing filament pyrometer has a range of about 600 °C to 3000 °C, an accuracy of about ±0.5% of the reading and involves no physical contact with the hot object. It can thus be used for moving or distant objects.

Corrections have to be made for the emissivity ε of the hot object. The true temperature T of the object is related to the apparent temperature T_a, based on the assumption that the object is a black body, by the relationship

$$\frac{1}{T} - \frac{1}{T_a} = \frac{\lambda \ln \varepsilon}{c_2}$$

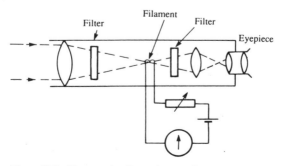

Figure 21.5 Disappearing filament pyrometer

where the temperatures are on the kelvin scale, λ is the wavelength at which the measurement is made (typically about 0.65 μm) and c_2 is a constant with the value 1.4388×10^{-2} m K.

10 Radiation pyrometer

With the radiation pyrometer the radiation from the object is focused onto a radiation detector (Figure 21.6). This might be a broad band detector such as a thermopile, resistance thermometer, or a thermistor. A *broad band detector* detects the radiation over a wide band of frequencies and so the output from the detector is the summation of the power emitted at every wavelength and is represented by the area under the Figure 21.4 graph at a particular temperature. Hence the detector output is proportional to the fourth power of the temperature in kelvin.

An alternative to a broad band detector is a *narrow band detector* or *photon detector* such as photoconductive and photoemissive cells. These are much quicker at responding than the broad band detectors, microseconds rather than milliseconds, and are responsive to only a narrow band of wavelengths, a filter sometimes being added to narrow the band yet more. The output of such a detector is proportional to

$$\frac{\Delta\lambda}{\lambda^5}\exp(-c_2/\lambda T)$$

where λ is the centre of the wavelength pass band, $\Delta\lambda$ the width of the band, c_2 a constant and T the temperature on the kelvin scale.

In some forms of instrument a rotary mechanical disc or shutter is used to chop the radiation before it reaches the detector and so give an alternating output, thermistors generally being the detector. Chopped signals can be amplified and so enable readings to be obtained where the level of radiation from the measured body is low.

The accuracy of radiation pyrometers is typically of the order of $\pm 0.5\%$ and ranges are available within the region 0 °C to 3000 °C, or higher. With broad band instruments the time constant varies from about 0.1 s with virtually just one thermocouple as detector to a few seconds with a thermopile involving many thermocouples. The time constant for narrow band instruments is typically of the order of a few microseconds. Chopped broad band instruments generally use thermistors since the time constant for thermopiles is too long. Chopped narrow band instruments, because of the small time constant, can use high chopping frequencies. Radiation pyrometers have the great advantage that they do not have to be in contact with the object whose temperature is being measured. They can thus be used for objects which are too hot for contact, too corrosive, or moving. For emissivity corrections see the previous item.

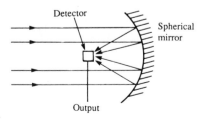

Figure 21.6 Radiation pyrometer

11 Two-colour pyrometer

The output of the radiation pyrometers described above is affected by changes in the emissivity of the hot object and changes in the transmission characteristics of the intervening medium between object and pyrometer and so calibration is necessary for a particular situation. These problems are overcome with the two-colour pyrometer (Figure 21.7). The incoming radiation from the object is split into two equal parts. Each part is then transmitted through a narrow band filter which allows only a narrow wavelength band of radiation through, the wavelengths being different for the two parts. The powers of the two bands of radiation are the monitored by detectors. The ratio of the outputs from these detectors is then used as a measure of the temperature of the object.

$$\text{Output ratio} = \left(\frac{\lambda_2}{\lambda_1}\right)^5 \exp \frac{c_2}{T}\left(\frac{1}{\lambda_2} - \frac{1}{\lambda_1}\right)$$

where λ_1 and λ_2 are the centre wavelengths of the two filters, T is the temperature on the kelvin scale and c_2 is a constant. The accuracy of such an instrument is typically of the order of $\pm 0.5\%$ with a range from about $0\,^\circ C$ to $3000\,^\circ C$, or higher. The result is independent of the emissivity of the object.

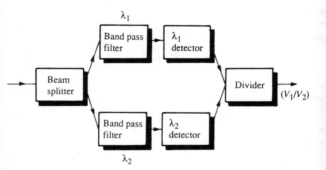

Figure 21.7 Two-colour pyrometer

22 Vacuum

The term vacuum is used for pressure below the atmospheric pressure (100 kPa), usually significantly below it. Some of the pressure measurement systems described in Chapter 18 can be used for pressures below atmospheric pressure. The U-tube manometer compares two pressures, and if one is the atmospheric pressure the other can be down to about 1 kPa. The Bourdon tube gauge can generally be used down to about 10 kPa. Diaphragm gauges can be used down to about 1 Pa. Table 22.1 lists the measurement systems discussed in this chapter which can be used for even lower pressures.

Further reading: Leck, J. H. (1964), *Pressure Measurement in Vacuum Systems*, Institute of Physics; Noltingk, B. E. (ed.) (1985), *Jones' Instrument Technology*, vol. 1 (*Mechanical Measurements*), Butterworth-Heinemann.

Boyle's law

For a fixed mass of ideal gas the pressure is inversely proportional to the volume. Thus by compressing a known volume of gas into a known smaller volume the pressure is increased by the ratio of these volumes and can be made large enough to become more easily measured.

1 Mcleod gauge

The Mcleod gauge (Figure 22.1) initially has the mercury level at A so that the gas in the bulb is at the pressure to be measured. The mercury level is then raised by allowing air to enter the mercury reservoir. When the mercury passes B the sample of the gas in the bulb is trapped. The mercury level is raised until it reaches C, the level in the open capillary tube corresponding to the top of the capillary tube of the trapped volume. The difference in heights h between the mercury in the two capillary tubes is then a measure of the gas pressure p. If V is the volume of gas trapped then it has been compressed into a volume

Table 22.1 Vacuum measurement systems

Principle	System	Characteristics
Boyle's law	1. Mcleod gauge	Range 5×10^{-4} to 10^5 Pa, accuracy ± 5 to 10%.
Thermal conductivity	2. Pirani gauge	Range 10^{-2} to 10^3 Pa, accuracy ± 10%.
	3. Thermistor gauge	Range 10^{-2} to 10^3 Pa, accuracy ± 10%.
	4. Thermocouple gauge	Range 10^{-1} to 10^3 Pa, accuracy ± 20%.
Ionization	5. Discharge tube	Rough indicator of stage of vacuum.
	6. Penning gauge	Range 10^{-5} to 1 Pa, accuracy ± 10 to 20%, shows hysteresis.
	7. Hot cathode gauge	Range 10^{-6} to 10^2 Pa, accuracy ± 10 to 30%.
	8. Bayard–Alpert gauge	Range 10^{-8} to 1 Pa, accuracy ± 10 to 30%.

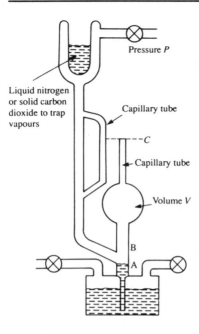

Figure 22.1 Mcleod gauge

hA, where A is the cross-sectional area of the capillary tube, by increasing the pressure by $h\rho g$. Thus using Boyle's law,

$$pV = (p + h\rho g)hA$$

where ρ is the density of the mercury and g the acceleration due to gravity. Hence

$$p = \frac{Ah^2\rho g}{V - Ah}$$

Since Ah is much smaller than V the equation approximates to

$$p = \frac{A\rho gh^2}{V}$$

The gauge can be used for the measurement of pressures in the range 5×10^{-4} Pa to atmospheric pressure with an accuracy of about ± 5 to 10%.

Thermal conductivity

The temperature of a heated wire or resistance element when a current passes through it depends on the rate at which heat is conducted away by the gas surrounding the element, and hence the gas pressure.

2 Pirani gauge

This gauge consists of a platinum or tungsten wire in a glass or metal tube. The electrical resistance of the wire depends on the pressure in the tube. Since it is also affected by changes in the ambient temperature

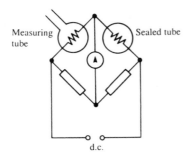

Figure 22.2 Wheatstone bridge with Pirani gauge

a dummy gauge is used, this being a sealed gauge for which the pressure does not change. The active and the dummy gauges are connected into adjacent arms of a Wheatstone bridge (Figure 22.2) and the out-of-balance current becomes a measure of the gas pressure in the active gauge tube. The Pirani gauge is used for pressure measurements in the range 10^{-2} to 10^3 Pa with an accuracy of about $\pm 10\%$.

3 Thermistor gauge
This gauge operates on the same principle as the Pirani gauge, using a thermistor instead of the metal wire. Because thermistors give much larger changes in resistance and are very small the gauge is more sensitive than the metal wire gauge and responds more quickly to changes in pressure. It is used for pressure measurements in the range 10^{-2} to 10^3 Pa with an accuracy of about $\pm 10\%$.

4 Thermocouple gauge
The thermocouple gauge consists of an electrically heated wire with the temperature of the wire being monitored by means of a thermocouple attached to it, the thermocouple output becoming a measure of the gas pressure. It is used for pressure measurements in the range 10^{-1} to 10^3 Pa with an accuracy of about $\pm 20\%$.

Ionization gauges

When a gas is bombarded with electrons ionization can occur and as a consequence a current can flow between two electrodes. The amount of ionization produced depends on the number of gas molecules and hence the gas pressure. Gauges are referred to as cold cathode or hot cathode depending on whether the electrons are produced by ions colliding with a cold cathode or whether they are produced as a result of heating the cathode.

5 Discharge tube
The discharge tube is a cold cathode ionization gauge. It consists of a glass tube with two sealed-in electrodes (Figure 22.3), a high voltage being applied between the two electrodes. This results in a luminous glow appearing in the tube, the colour of the glow depending on the gases present. The glow occurs for pressures between about 10^3 Pa and 10^{-1} Pa, the upper limit being a consequence of electrons colliding too frequently with gas molecules and not being able to be accelerated to high enough energies to ionize them, while the lower limit is because the electrons travel the full length of the tube without meeting gas molecules. The gauge is generally just used as a rough indicator of the state of the vacuum.

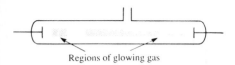

Regions of glowing gas

Figure 22.3 Discharge tube

6 Penning gauge

The Penning gauge is a cold cathode ionization gauge. It consists of two parallel flat cathodes, about 20 mm apart, in a glass or metal envelope (Figure 22.4). Midway between them is a wire anode. The gauge is placed between the poles of a magnet, either a permanent magnet or an electromagnet, which produces a magnetic field at right angles to the cathode plates. Electrons emitted from the cathodes move towards the anode in helical paths rather than straight lines. The consequence of this is that the distances travelled by the electrons between the electrodes is considerably increased and so the probability of a collision between an electron and a gas molecule is increased. A greater amount of ionization means a greater current between the electrodes. The current between the electrodes is taken as a measure of the gas pressure in the gauge, giving a reasonably linear relationship. The Penning gauge is used for pressures in the region 10^{-5} to 1 Pa with an accuracy of about ± 10 to 20%. A problem with the gauge is that it shows a hysteresis effect, the current depending on whether the pressure is increasing or decreasing.

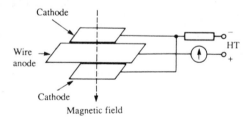

Figure 22.4 Penning gauge

7 Hot cathode gauge

This has electrons emitted by a heated filament. Surrounding the filament is a grid for the collection of electrons and surrounding that the cylindrical ion-collector electrode (Figure 22.5). The grid is positive with respect to the filament and collects the electrons produced as a result of collisions between the electrons emitted from the filament and gas molecules. The outer electrode is negative with respect to the filament and the grid and attracts the positive ions. The pressure P is proportional to the ratio of the positive ion current i_+ and electron current i_-. Thus

$$i_+ = kPi_-$$

where k is a constant called the gauge factor. This constant is usually

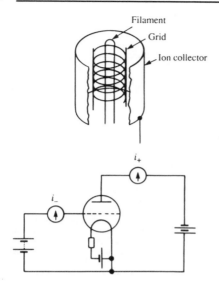

Figure 22.5 Hot cathode ionization gauge

quoted for when the gas is nitrogen. For other gases it is multiplied by a relative sensitivity factor C, to account for the probabilities of ions being produced. If the electron current i_- is kept constant then the positive ion current i_+ is directly proportional to the pressure. This gauge is used for pressures in the range 10^{-6} to 10^2 Pa with an accuracy of about ± 10 to 30%.

8 Bayard–Alpert gauge

The lowest pressure measurable with the hot cathode gauge is determined by the X-rays produced as a result of the electrons striking the grid. The X-rays are absorbed by the ion-collector and in doing so produce photoelectrons which are then collected by the grid and give rise to an extra electron current which is superimposed on that current which is related to the pressure. This current is of the order of that which is produced when the pressure is 10^{-6} Pa. This effect is reduced by a modification of the design by Bayard and Alpert (Figure 22.6) in which the ion collector is a fine wire and so a much smaller surface area for the interception of the X-rays. Such a gauge can be used for the measurement of pressures from about 1 Pa down to 10^{-8} Pa with an accuracy of about ± 10 to 30%.

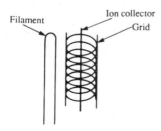

Figure 22.6 Bayard–Alpert ionization gauge

Part Four
Microprocessor based systems

23 Microprocessors in instrumentation

Intelligent instruments

The term *intelligent* is applied to measurement systems if they include a microprocessor to carry out the main elements of signal processing. With a conventional *dumb* instrument, the system might involve more than one sensor, each with the appropriate signal converter, to give a number of measurements which then have to be combined by the operator to give the value of the measured quantity, e.g. humidity measurement where the 'wet' and 'dry' temperatures have to be measured, or where the measurement has to be corrected for, perhaps, non-linearity. Such dumb instruments can thus involve, for example, arithmetic involving computations with a number of measurements, or perhaps looking up calibration data or incorporating correction factors for non-linearity. The operator is thus an element in the signal processing needed to give the measured value. A microprocessor system can be used to replace the operator so that the resulting intelligent instrument gives the measured value directly from, perhaps, simultaneous inputs from a number of sensors or perhaps from a number of measurements made at different times by a single sensor. In addition, a microprocessor based instrument can carry out a number of other tasks, e.g. convert data into a different format, take the average of a data, display the maximum and minimum values of data, carry out a sequence of operations such as sampling a number of different transducers and periodic calibration, make decisions based on measurements for control systems, etc.

As an illustration, Figure 23.1 shows the type of system that might be employed with a weighing machine based on the use of a load cell when the instrument is perhaps in the checkout of a supermarket and is used to give a display of the cost of a bag of apples or some other produce. It might also give feedback to the warehouse of the amount sold so that stock levels can be monitored.

A microprocessor controlled weighing machine can have self calibration built into it. When it is commanded to calibrate, a standard weight is automatically placed on the load cell and the weight found. This value can then be compared with the correct weight that is stored in the memory of the system and any difference used to automatically correct future weighings.

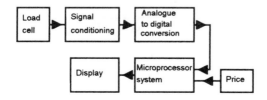

Figure 23.1 Microprocessor-based weighing machine

As another example, Figure 23.2 shows the elements of a microprocessor controlled thermometer employing a thermocouple. With a thermocouple the e.m.f. is not a linear function of the temperature and tables have to be used to convert e.m.f. values into temperatures. Also, the tables assume that the cold junction of the thermocouple is at 0°C. If it is not then corrections have to be made. The microprocessor controlled thermometer thus has the thermocouple hot junction to monitor the temperature being measured and a resistance element to monitor the temperature of the thermocouple cold junction. After signal conditioning, the signals from the two elements are fed, in turn, to the microprocessor system which then computes the value of the temperature.

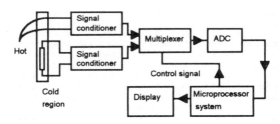

Figure 23.2 Microprocessor controlled thermometer

The measurement of relative humidity requires the measurement of the temperature of a dry thermometer and the measurement of the temperature of a wet thermometer. With a manual form of instrument, the two values are then looked up in tables and the relative humidity obtained. An instrument which directly gives the relative humidity can be constructed using a microprocessor. Figure 23.3 illustrates the system. The temperature sensors might be piezoelectric, e.g. quartz crystals, and have a frequency which is temperature dependent. The microprocessor system thus has two inputs from which it is able to compute, using standard tables of values, the relative humidity. The results displayed might be the two temperatures and the relative humidity.

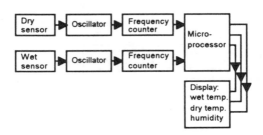

Figure 23.3 Relative humidity measurement

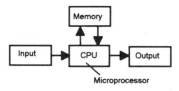

Figure 23.4 Microprocessor system

Basic elements of microprocessor systems

The basic elements of a microprocessor system (Figure 23.4) are the *central processing unit* (CPU), *memory* and *input and output interfaces*.

The microprocessor is the central processing unit (CPU) and is responsible for executing arithmetic and logic operations on binary data and also for controlling the timing and sequencing of operations in the system. There are two main forms of memory. *Read-only memory* (ROM) is a permanent data store with data entered into it at the manufacturing stage. The CPU can only read data from ROM, being unable to enter (the term write is used) data into it. ROM is used to contain the microcomputer operating system and standard routines that might be used frequently. The second type of memory is *random access memory* (RAM). Data is stored temporarily in this memory with the CPU able to write new data into it. Unless a back up battery is used, the data stored in RAM is lost when the power supply is turned off. RAM is used to store the user's programs. The input and output units are the interfaces between the microprocessor and external devices, the terms input port and output port are used for such interfaces. The input port receives signals from external devices; these must always be in binary form when they reach the CPU. The output port receives binary signals from the CPU for transmission to external devices.

The main elements in a microprocessor system are connected together by what are termed buses (Figure 23.5). A *bus* is a group of connecting tracks or wires along which digital signals are sent.

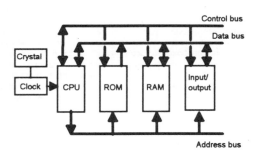

Figure 23.5 Microcomputer architecture

There are three main buses in a microprocessor . The *data bus* is used to transfer data between the microprocessor and the other elements in the system. The *address bus* is used to send the addresses of memory locations so that data can be retrieved or stored. The *control bus* is used to send control signals from the microprocessor to the other elements in the system.

The buses transmit signals by what is termed *parallel transfer*. *Serial transfer* would involve each bit in turn being sent along a single wire. For example, with parallel transfer the binary value 0101 1011 representing the value of some input signal has each of its bits simultaneously sent along a separate wire in the bus. The number of input lines on which data can enter a microprocessor is determined by the width of the data bus and microprocessors are categorised according width of the data path, e.g. an 8-bit microprocessor. The term *word* is used for the number of bits used to contain a piece of information. Word lengths might be 4, 8, 16, 32 or 64 bits. The word length determines the number of values that can be represented by the data. With a word length of 4 bits, the number of values possible is $2^4 = 16$, with an 8-bit word $2^8 = 256$ and with a 16-bit word $2^{16} = 65\ 536$. Thus suppose we have a temperature sensor with a range of 0°C to 120°C and we need to be able to recognise temperature changes of, say, 0.5°C. The microprocessor thus needs to be able to give different binary values to each of 240 different input signals. Thus an 8-bit microprocessor would have the required width of data path for such a system.

The microprocessor

Basically, the microprocessor (Figure 23.6), i.e. CPU, contains an arithmetical and logical unit, a control unit and registers, i.e. memory locations, in which information involved in program execution can be stored.

1   The *arithmetical and logical unit* (ALU) communicates with the other units, receiving data from the input unit and sending data out via the output unit, and carrying out operations in accordance with the set of instructions, termed the program, which is stored in the memory. It can also send data for storage into memory.

2   The entire process is synchronised by the *control unit* which is responsible for the timing of each instruction and the flow of data. The number and types of register vary from one microprocessor to another.

3   Basic registers are the *accumulator* where data for an input to the ALU is temporarily stored, the *flag register* which stores information concerning the result of the latest process carried out in the ALU, the *program counter* which allows the microprocessor to keep track of its position in a program, the *stack pointer* in which program counter values can be stored to enable particular parts of a program to be accessed, the *instruction and decoder register* which stores an instruction for decoding and subsequent execution.

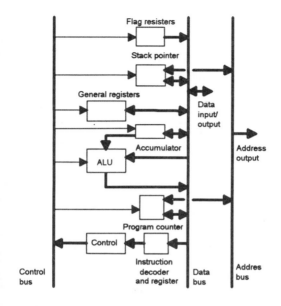

Figure 23.6 Basic elements of a microprocessor

Microprocessors in general need interface devices between their inputs/outputs and the peripheral devices to which they are connected. This is because the signals originating from a peripheral device are often different from those which the microprocessor requires or are at a different speed to that which the microprocessor can handle. Likewise, the signals required by a peripheral device my be different or at a different speed to those emanating from a microprocessor. See Chapter 24 for a discussion of interfaces.

Microcontrollers

Microcontrollers are integrated systems on a single chip involving a microprocessor with memory and input/output interfaces. Figure 23.7 shows a general block diagram of a microcontroller. They typically have a number of input and output ports, with often one or more ports being able to be programmed to be either input or output ports. One input port might also be supplied with its own analogue-to-digital converter and so enable analogue inputs to be directly connected to the port. Often some of the channels in a port may be able to be programmed to accept or transmit serial signals. Other devices such as timers are often included.

Further general reading on microprocessors: Anderson, J.S. (1994) *Microprocessor Technology,* Butterworth-Heinemann; Bolton, W. (2000) *Microprocessor Systems,* Longman; Vears, R. (1996) *Microelectronic Systems,* Butterworth-Heinemann

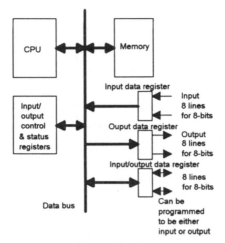

Figure 23.7 Basic elements of a microcontroller

Basic elements of a microprocessor based instrument

Figure 23.8 shows the basic elements of a digital thermometer. The sensor is a thermotransistor which, together with signal conditioning, is available as an integrated package and gives an output which is proportional to the temperature (National Semiconductor LM35 gives 10 mV/°C over the range −40°C to 110°C). This gives an analogue signal which is the input to a port on a microcontroller which is able to accept analogue signals. The output from the microcontroller is taken as a serial signal which is fed via an encoder/driver to transform the signal into a suitable form to give a digital display.

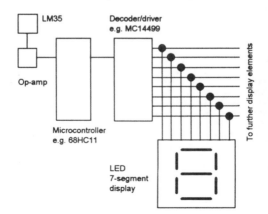

Figure 23.8 Digital thermometer

Figure 23.9 shows the general form of a typical microprocessor based instrument. The program to be followed is entered into the system via a keyboard or from a floppy disk and is stored in RAM. The basic system can be supplied with a number of different programs, depending on the tasks it is required to do. The ROM contains the setting up and subroutine programs which cannot be changed and are permanently stored in this memory by the chip manufacturer.

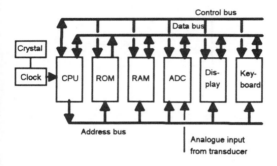

Figure 23.9 A microprocessor based instrument

The output from the transducer and its signal conditioning elements is likely to be analogue. When the program is run the microprocessor instructs the analogue to digital converter to take a sample of the analogue input from the transducer, convert it into a digital signal, e.g. an 8-bit parallel word, and place it in the data bus. The CPU receives this data and processes it in the manner to which it is instructed by the program. The result may then be displayed on perhaps a visual display unit (VDU), stored in RAM for future use or perhaps compared with a reference value stored in RAM and the result displayed or some action initiated. The entire sequence of events is then repeated with the microprocessor instructing the analogue to digital converter to take another sample of the analogue signal from the transducer.

Further reading: Bolton, W. (2000), *Microprocessor Systems*, Longman; Fraser, C.J. and Milne, J.S (1990), *Microcomputer Applications in Measurement Systems*, Macmillan; Tooley, M. (1990, 1995), *PC-based Instrumentation and Control*, Butterworth-Heinemann.

Data acquisition systems

There are often situations where it is necessary to determine simultaneously the values of several variables. This might, for example, be a measurement of flow rates, temperatures, pressures, etc. at various points in a chemical plant. Complex systems may receive data from hundreds of transducers. Rather than use a separate system for each measurement, a data acquisition (DAQ) system is used. Figure 23.10 shows the basic form of a data acquisition system.

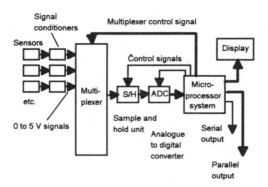

Figure 23.10 Basic elements of a data acquisition system

The outputs from each sensor pass through signal conditioning to perhaps convert them into a voltage, amplify them, linearize them, etc. in order to bring them to a common signal range, usually 0 to 5 V. The resulting voltage signals are then inputs to a multiplexer. The multiplexer is controlled by the microprocessor system so that it selects input signals. The selected signal is then passed via a sample and hold unit to an analogue to digital converter and so becomes a parallel word input to the microcomputer where it can be processed before being displayed or transmitted to remote displays or used in control systems or stored.

DAQ boards

If a PC is used, a basic system might have the multiplexer and analogue to digital converter on a board which is mounted in one of the expansion slots within the computer or, if the measurement needs exceed the capability of a plug in card, as a separate unit. The unit might include some signal conditioning elements that can be used with common sensors such as thermocouples. A typical card might have 16 analogue to digital input channels with 12 bit resolution and a conversion time of 25 μs. Figure 23.11 shows the basic elements of such a DAQ card.

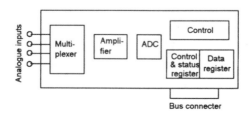

Figure 23.11 DAQ card

The specifications for DAQ boards for analogue inputs include such items as:

1 Sampling rate for analogue inputs
  This specifies the number of samples the analogue-to-digital converter takes per second.
2 Number of input channels
  Multiplexing is used to sample each channel in turn.
3 Resolution
  This is often specified as the number of bits that the analogue-to-digital converter uses to represent the analogue signal, e.g. 16-bit.
4 Range
  This is a specification of the minimum and maximum analogue voltages that the analogue-to-digital converter can handle, e.g. 0 to 10 V.
5 Gain
  The gain available from the amplifier on the board is generally selectable and this, coupled with the range and resolution, determines the smallest detectable change in input analogue voltage. Thus if the gain is 20 and the range 0 to 10 V for a 16-bit board, the smallest detectable input voltage is $10/(20 \times 2^{16}) = 7.6\ \mu V$.

DAQ board specifications for digital inputs include:

1 Number of channels
2 Digital logic levels
  This specifies the maximum and minimum voltages.
3 Handshaking
  In transferring digital data between a computer and external equipment, the circuitry that has to be used for communication and synchronisation of communications (see Chapter 24) has to be specified.

Programming

The microcomputer has to be programmed to carry out the data acquisition. This might be done by writing a program in BASIC or some other language, or using software specially written to enable the program to be specified by the operator by selection from screen displays without having to write a program. With writing a program the procedure to be defined is:

1 Define the card base address.
2 Select an input channel.
3 Send out a start conversion signal.
4 Check for the end of the conversion.
5 Read the output from the analogue to digital converter.
6 Store the data in memory.
7 Various looping routines can be devised to monitor continuously one particular input or scan through channels by, perhaps, repeating steps 2 to 6 for other channels.
8 The result might then be printed out or otherwise displayed or further processed.

There is a range of software commercially available which can be used to control the computer when it is used for data acquisition, avoiding the need to actually write a program, and present the output on a visual display unit. The software also has an application editor which permits the operator to specify how the data sources and processing functions are to be handled in a particular application. This is commonly achieved by the operator filling in pop-up panels and/or manipulating slider controls on the computer screen and/or wiring together boxes. Some software has the ability to carry out a high degree of data manipulation and display. There is also likely to be the facility to output collected or processed data in ASCII format, for processing by other software. Examples of such software are: DT VEE, Lab Tech Notebook$_{pro}$, Lab View, Lab Windows, Signal Centre Professional , Test point and Virtual Designer.

Further reading: Paton, B.E. (1999) *Sensors, Transducers & LabView*, Prentice Hall

Data logger

The term *data logger* is used for a unit that might be added to a measurement system. It is programmed to capture and store measurements from a range of transducers. At a later time they might be connected to a microcomputer to download the captured data for analysis and display. The data logger may be programmed independently of a computer or via a link to a computer running suitable software. A data logger is useful for remote, unsupervised monitoring applications.

24 Interfacing

Standard bus

A microprocessor based system will often involve interconnections between various systems or subsystems. In making such connections it is necessary to ensure compatibility of such matters as connectors, control signals and logic levels. The connection process is thus generally simplified by using a *standard bus*. This specifies the type of connectors to be used, the signals available and their location on the connectors. Such a standard enables elements to be easily added to a system.

With digital communication of signals the signal transfer can be by either parallel or serial means. *Parallel communication* involves using a cable system with each data bit of a data word being sent simultaneously along parallel data lines. *Serial communication* involves data being sent in a chain-like manner along a single pathway. Parallel communication is a faster method of communication than serial but is restricted to a maximum transmission rate of 1 M bytes/s up to a maximum transmission distance of 15 m and is intended for high speed, short distance, communication in situations where there is little electrical interference. Serial data communication can be used over much longer distances.

General-purpose instrument bus

The standard bus which is most commonly used with instrumentation systems for parallel data transfer is the *IEEE-488 system bus* (also known as the *GPIB, general purpose interface bus*). This bus was originally developed by Hewlett Packard and was originally called the *Hewlett Packard instrument bus* (HPIB). There is a total of 24 lines, of which eight are ground return lines, and three sets of bus lines (Figure 24.1). One set is the data bus. This consists of eight bi-directional lines and is used to carry data and commands between the various devices connected to the bus. Another set is the management bus which has five lines to give control and status signals. The remaining three lines are used for handshaking between the devices on the bus.

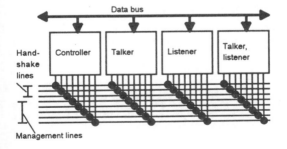

Figure 24.1 IEEE-488 bus structure. Talkers are devices sending out data, listeners are devices receiving data.

Devices connected to the bus may act as either talkers or listeners. Talkers send data to the bus and listeners receive data. Only one device can act as a talker at any time though several devices may be listeners. One of the devices on the bus acts as the bus controller. Devices can be switched from being listeners to talkers by sending suitable commands along the data lines. Each device on the bus can have its own address and up to 31 devices can be addressed. Devices addresses are sent via the data bus as a parallel 7-bit word, the lowest 5-bits providing the device address and the other two bits controlling information. If both these bits are 0 then commands are sent to all addresses, when bit 6 is 1 and bit 7 is 0 the addressed device is switched to be a listener, if bit 6 is 0 and bit 7 is 1 it is a talker.

The management lines each have an individual task in the control of the flow of information across the interface between systems or subsystems. The following are the five management lines:

1   IFC (interface clear)
    This is used by the controller to reset all devices of the system to the start state.
2   ATN (attention)
    This is used by the controller when inserting a device command on the bus. If the level is low on this line then there is a device command, if high then the signal is to be interpreted as data rather than a command.
3   SRQ (service request)
    This is used by devices to signal the controller that they need attention.
4   REN (remote enable)
    This enables a device on the bus to indicate that it is to be selected for remote control rather than by its own control panel.
5   EOI (end or identify)
    This is used to either signify the end of a message sequence from a talker device or is used by the controller to ask a device to identify itself.

When a device has been selected as the talker, it can send data along the bus to all of the devices that have been set up as listeners. This is done by the device setting the SRQ line low to request service. The controller then sends the ATN signal and the talker can send data to the bus.

The handshake lines are used for controlling the transfer of data. These three lines ensure that the talker will only talk when it is being listened to by listeners.

1   DAV (Data valid)
    When the level is low on this line then the information on the data bus is valid and acceptable.
2   NRFD (Not ready for data)
    This line is used by listener devices, using high level, to indicate that they are ready to accept data.
3   NDAC (not data accepted)
    This line is used to listener devices, using high level, to indicate the acceptability of data.

Initially DAV is high indicating that there is no valid data on the data bus, NRFD and NDAC being low. A data word might then be put on the data bus. Then the NRFD line goes high to indicate that all the listeners are ready to accept data. DAV then goes low to indicate that new data has been placed on the data bus. When a device accepts a data word it sets NDAC to high to indicate that it has accepted the data and NFRD to low to indicate that it is now not ready to accept data . When all the listeners have set NDAC to high then the talker cancels the data valid signal, DAV going high. This then results in NDAC being set low. The entire process can then be repeated by another word being put on the data bus. Figure 24.2 illustrates this sequence.

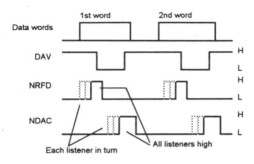

Figure 24.2 Handshaking sequence

The controller sends multi-line commands over the data bus as data bytes with ATN set high. Commands, such as ATN, which involve only a single line are termed *uni-line messages*, messages characterised by the state of several lines are termed *multi-line messages*. Multi-line messages fall into five groups: commands which are addressed to listeners to select bus functions (ACQ), universal commands to select bus functions affecting all devices (UCQ), commands concerning listeners' addresses (LAG to set a specified device to listen, UNL to set all devices to unlisten status), commands concerning talkers' addresses (TAG to set a specified device to talk, UNL to set all devices to untalk status), and secondary commands (SCQ) used to specify a device sub-address or subfunction. As an illustration, the command TAG is sent by having DI07 set to high, DI06 to 0, ATN to 1, and the setting of the other inputs does not affect the outcome.

In order to make use of the IEEE-488 bus it is necessary for the computer to have a software driver installed which will act as an interface to software applications packages. The applications packages are then able to access the facilities offered by the bus using commands such as ENTER, OUTPUT, etc.

The IEEE-488 bus generally uses a standard 24 pin connector to make connections with devices. The pin allocations are given in Table 24.1.

Table 24.1 IEEE-488 bus system

Pin	Signal group		Function
1	Data	D101	Data line 1
2	Data	D102	Data line 2
3	Data	D103	Data line 3
4	Data	D104	Data line 4
5	Management	EOI	End Or Identify: used either to signify the end of a message sequence from a talker or by the controller to ask a device to identify itself
6	Handshake	DAV	Data Valid: when low then the information on the data bus is valid and acceptable
7	Handshake	NRFD	Not Ready For Data: used by listener devices taking it high to indicate that they are ready to accept data
8	Handshake	NDAC	Not Data Accepted: used by listeners taking it high to indicate that data is being accepted
9	Management	IFC	Interface Clear: used by the controller to reset all the devices of the system to the start state
10	Management	SRQ	Service Request: used by devices to signal to the controller that they need attention
11	Management	ATN	Attention: used by the controller to signal that it is placing a command on the data lines
12		SHIELD	Shield
13	Data	D105	Data line 5
14.	Data	D106	Data line 6
15	Data	D107	Data line 7
16	Data	D108	Data line 8
17	Management	REN	Remote Enable: to enable a device to indicate that it is to be selected for remote control rather than by its own control panel
18		GND	Ground/common
19		GND	Ground/common
20		GND	Ground/common
21		GND	Ground/common
22		GND	Ground/common
23		GND	Ground/common
24		GND	Signal ground

Centronics parallel interface

The Centronics parallel interface is commonly used for the parallel interface to a printer. Table 24.2 shows the pin assignments. The signal levels used for the Centronics interface are transistor–transistor logic (TTL).

Table 24.2 Centronics pin assignments

Signal pin	Return pin	Signal	Function
1	19	STROBE	Strobe pulse to read data in
2	20	DATA 1	Data bit 1 (LSB)
3	21	DATA 2	Data bit 2
4	22	DATA 3	Data bit 3
5	23	DATA 4	Data bit 4
6	24	DATA 5	Data bit 5
7	25	DATA 6	Data bit 6
8	26	DATA 7	Data bit 7
9	27	DATA 8	Data bit 8 (MSB)
10	28	ACK	Acknowledge pulse to indicate data has been received and printer ready for new data
11	29	BUSY	Printer busy, a high signal indicating printer cannot receive data
12		PO	Goes high when printer out of paper
13		SLCT	Select status, high when printer can communicate, low when not
14		AUTO FEED	Auto linefeed, if low a line-feed is added to a carriage return
16		SG	Signal ground
17		FG	Frame ground
18		+5 V	
31	30	PRIME	Used to initialise printer, when low the printer resets
32		ERROR	Error status line, low when printer detects a fault
33		SG	Signal ground

Figure 24.3 shows how a printer can be connected via a Centronics bus to a Motorola 68HC11 microcontroller. In the illustration, port C of the microcontroller is set up for the output of data to the printer and the microcontroller sends a strobe pulse from its STRB pin to the printer every time it sends data to the printer. Handshaking involves two printer signals, ACK and BUSY. When the printer receives the strobe pulse it sets its BUSY line high; and after the data has been received it sends back an acknowledge pulse, indicating it is ready for more data, and sets BUSY low. The acknowledgement of receipt of signal

ACK by the printer gives an input to the microcontroller STRA pin; the BUSY signal is not used in the example illustrated by the figure. The select state SLCT signal is provided by an input to pin PE7 of the microcontroller; a low indicates that the printer is effectively disconnected from the microcontroller and a high indicates that the printer can communicate with the microcontroller. The error signal ERROR is inputted to pin PE6 of the microcntroller and goes low when the printer detects a fault.

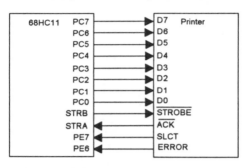

Figure 24.3 Interfacing a microcontroller to a Centronics printer

Serial data transmission

Digital serial data transmission links may be divided into three categories:

1   *Simplex*
    This involves communication from A to B where B is not capable of transmitting back to A and uses two wires.
2   *Half duplex*
    This involves transmissions from A to B and B to A but not simultaneously and involves a two or four wire cable.
3   *Full duplex*
    This involves simultaneous transmissions from A to B and B to A. This involves a two or four wire cable.

With each of the above forms of communication, it is necessary for a receiver to be ready to receive and identify each set of data as it is received from the transmitter. This can be done in two possible ways. *Asynchronous transmission* involves each frame of data being preceded by a start bit and terminated by a stop bit. The receiving device then knows exactly where data starts and ends. Because of the need to check for start and stop bits, such a form of transmission is restricted to less than 1200 bits/s and tends to be used for intermittent transmission. With *synchronous transmission* there is no need for start and stop bits since the transmitter and receiver are synchronised to a common clock. The transmitting data is preceded by a synchronizing clock signal and then each character frame is recognised as a block of 7 or 8 bits. Periodic resynchronisation occurs. This mode of transmission can be used with more than 1200 bits/s and is commonly used where there is an almost constant stream of data, e.g. computer files.

A commonly used standard serial interface is the *RS-232*. This interface can be used for transmission rates up to 20 000 bits/s over distances up to about 15 m with lower bit rates at longer distances. The interface RS-449 is a later standard and improves on the bit rate and distance, bit rates of 10 000 bits/s over a distance of 1 km being feasible. The RS-232 uses +12 V for logic 0 and −12 V for logic 1. Microprocessors, however, use transistor-transistor logic (TTL) with logic 0 as 0 V and logic 1 as +5 V. Signal levels have thus to be converted, This can be achieved by using integrated circuits such as MC1488 for TTL to RS-232 conversion and MC1489 for RS-232C to TTL conversion.

Table 24.3 shows the pin assignments for RS-232; the signals can be grouped into three categories:

1 *Data*
  RS-232 provides for two independent serial data channels, termed primary and secondary. Both these channels can be used for full duplex operation.
2 *Handshake control*
  Handshaking signals are used to control the flow of serial data over the communication path.
3 *Timing*
  For synchronous operation it is necessary to pass clock signals between transmitters and receivers.

Table 24.3 RS-232 pin assignments

Pin		Direction	Signal/function
1	FG		Frame/protective ground
2	TXD	To DCE	Transmitted data
3	RXD	To DTE	Received data
4	RTS	To DCE	Request to send
5	CTS	To DTE	Clear to send
6	DSR	To DTE	DCE ready
7	SG		Signal ground/common return
8	DCD	To DTE	Received line detector
12	SDCD	To DTE	Secondary received line signal detector
13	SCTS	To DTE	Secondary clear to send
14	STD	To DCE	Secondary transmitted data
15	TC	To DTE	Transmit signal timing
16	SRD	To DTE	Secondary received data
17	RC	To DTE	Received signal timing
18		To DCE	Local loop back
19	SRTS	To DCE	Secondary request to send
20	DTR	To DCE	Data terminal ready
21	SQ	To DEC/DTE	Remote loop back/signal quality detector
22	RI	To DTE	Ring indicator
23		To DEC/DTE	Data signal rate selector
24	TC	To DCE	Transmit signal timing
25		To DTE	Test mode

For the simplest bi-directional link only the two lines 2 and 3 for transmitted data and received data, with signal ground and handshaking lines, are required (Figure 24.4). Data termination equipment (DTE) is normally fitted with a male connector and data circuit terminating equipment (DCE) with a female connector. Data terminating equipment, e.g. a microcomputer, is capable of sending and/or receiving data, data circuit terminating equipment is generally a device which can facilitate data communication, e.g. a modem. While data transmitted over such an interface can use any convenient code, ASCII is the one commonly used for computer systems.

Further reading: Petersen, D. (1992), *Audio, Video and Data Telecommunications*, McGraw-Hill; Green, D.C. (1991, 1995), *Data Communication*, Longman.

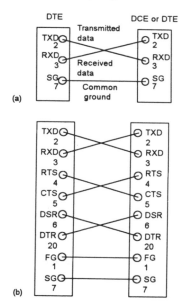

Figure 24.4 Connections: a) minimum, (b) more complex

I²C bus

The *inter-IC communication bus*, referred to as the I²C bus, is a data bus designed by Philips for use in communications between integrated circuits on the same circuit board or between equipment when the joining cable is relatively short. It allows data and instructions to be exchanged between devices by means of just two wires. The two lines are a bi-directional serial data line (SDA) and a serial clock line (SCL) and both lines are connected to the positive power supply via resistors (Figure 24.5). The device that controls the bus operation is the master, and the devices it controls are the slaves. Some microcontrollers have I²C interfaces.

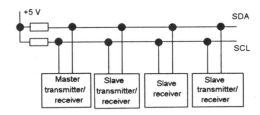

Figure 24.5 I²C bus

When the bus is in the idle state with no input, both the clock and the data lines are high (Figure 24.6). To initiate a data transfer, the transmitter pulls down the SDA bus followed by the SCL bus line. This gives the start signal and data is then sent. Receipt, and so the end of the data byte, is indicated by an acknowledge signal; the receiver takes the SCL line high then the data line SDA high. There is one clock pulse per data bit transferred with no limit on the number of data bytes that can be transferred between the start and stop conditions; after each byte of data the receiver acknowledges with a ninth bit. The next byte can then be transmitted. When the end of the message is reached, a stop signal is sent by taking SCL high then SDA high.

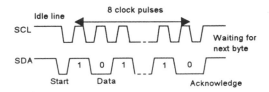

Figure 24.6 Bus conditions

Interfacing peripherals

Microprocessor systems are designed to operate with inputs of the order of a few volts and small currents and are damaged if higher voltages or currents occur at their inputs or outputs. The input might be from a sensor giving a digital output or from an analogue to digital converter as part of a sensor-signal conditioner arrangement. These might be giving higher voltage digital signals than the microprocessor can withstand or be subject to high transient voltage pulses or noise. Likewise, equipment to be operated by the microprocessor might require different voltages and/or currents than the microprocessor outputs supply. There is also the problem that the microprocessor might operate at different bit rates, and handshaking lines to control the timing of data transfers can be required. The following illustrates how some of the above problems can be overcome.

Electrical isolation

Because of different voltages/currents, there is often a need to electrically isolate input/output devices from the microprocessor inputs/outputs. A common form of isolating device used is the *opto-isolator or optocoupler*. Figure 24.7 shows the basic form of such a device with the appropriate circuit for connecting it to, on the input side, the digital input from a sensor with suitable signal conditioning and, in the output side, a microprocessor. It consists of a light-emitting diode (LED) through which the input current flows. Infrared radiation is emitted by the LED and detected by a phototransistor in close proximity but electrically isolated from it. Thus pulses in the LED circuit produce pulses in the photo- transistor circuit, without there being any physical connection between them.

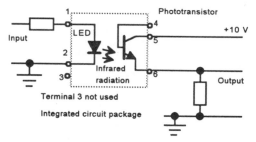

Figure 24.7 Opto-isolator

The above shows a single opto-isolator; integrated circuit packages are, however, available with two or four opto-isolators in the same package. The *transfer ratio* is used to specify the ratio of the output current to the input current. For example, the single transistor opto-isolator shown in Figure 24.7 might have a transfer ratio of 20% and an output current limited to 7 mA.

There are also forms incorporating other components and designed to operate in particular situations (Figure 24.8). Opto-isolators are available with built-in diodes to give rectification and so use with a.c. inputs (Figure 24.8(a)). Darlington transistor pairs (Figure 24.7(b)) are used for high gain, such an opto-isolator having perhaps a transfer ratio of 300% and a maximum output current of 100 mA. Built-in triacs and silicon controlled rectifiers (SCRs) are available for when an opto-isolator is used in the output line to drive higher power circuits. The triac conducts on either half of an a.c. cycle and thus can be used with a.c. circuits. The triac opto-isolator with the zero crossing unit turns on the triac only when the voltage goes through zero and consequently reduces the on transients and prevents electromagnetic interference.

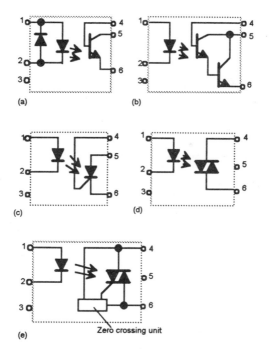

Figure 24.8 Opto-isolator types, (a) ac input-transistor output, (b) Darlington, (c) SCR, (d) triac, (e) triac with zero crossing circuit

Interfacing power

Opto-isolators can be used to drive directly low power load circuits. With higher power loads they are likely to be used with a relay or transistor switch circuit.

Relays can be used in conjunction with opto-isolators to enable their relatively small current output to be used to switch on much larger currents. Thus a relay might be used to enable the output from a microprocessor to operate a relay and switch on or off a motor. An alternative is to use a transistor switch. Figure 24.9 shows a basic transistor switch. The transistor is switched on or off by means of the signal applied to its base. By using the output from the microprocessor to repeatedly switch the motor on and off and vary the fraction of the time for which the motor is switched on, the average current applied to the motor can be varied and so its speed. The circuit in Figure 24.9 can only drive the motor in one direction. Figure 24.10 shows a transistor switch circuit employing four transistors; this can be used to operate the motor in both forward and reverse directions and control the speed in both directions.

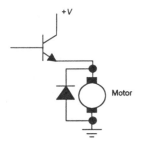

Figure 24.9 Motor speed control with a transistor switch

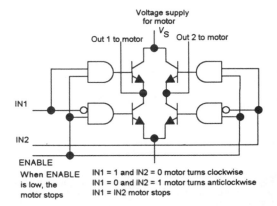

Figure 24.10 Motor speed and direction control

When ENABLE is low, the motor stops

IN1 = 1 and IN2 = 0 motor turns clockwise
IN1 = 0 and IN2 = 1 motor turns anticlockwise
IN1 = IN2 motor stops

Buffers

If the output from a microprocessor is connected directly to the base of a transistor, the base current required to switch the transistor is likely to be higher than the current which can be supplied by the microprocessor. A buffer is thus used to step up the current and also provide isolation. The term *three-state* is used for a buffer which can have an output which can be low or high and also a state where it the buffer acquires a high impedance and so appears like an open-circuit. The high impedance state is controlled by an enable input to the buffer. Figure 24.11 shows three-state buffer symbols and their operating characteristics.

Interfacing LED displays

A light-emitting diode (LED) will be illuminated if it is forward biased and there is enough current flowing through it, typically a few milliamps. Seven-segment displays contain seven LEDs in a figure-eight arrangement to enable different characters to be formed by illuminating the relevant LEDs.

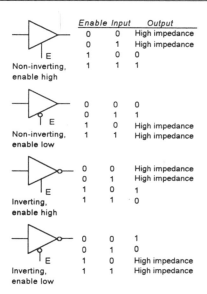

Figure 24.11 Three-state buffers

The seven-segment display requires inputs of seven on-off signals. Thus to display the required character the correct arrangement of bits has to be supplied by the microprocessor. The software used with the microprocessor has thus to take the program data and compute the required arrangement of the seven bits. This is then applied to a latch such as 74LS244; this is a circuit which holds the output to activate the LEDs (Figure 24.12). An alternative to using software is to use hardware to convert the microprocessor output to the required seven bits and also provide enough current to drive the LEDs. Figure 24.13 shows a system employing the decoder 7447.

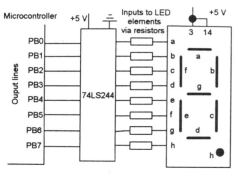

Figure 24.12 Latch system

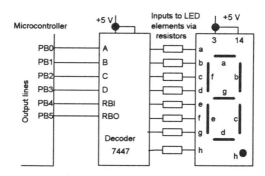

Figure 24.13 Decoder system

With displays having many display elements, rather than use a decoder for each element, multiplexing is used with a single decoder. Figure 24.14 illustrates such a system. Data is outputted from port A of a microcontroller and the decoder presents the decoder output to all the displays. Each display is connected to ground through an npn transistor such as the 2N2222. The display cannot light up unless the transistor is switched on by an output from port B. Thus by switching between the output lines from port B the output from port A can be switched to the appropriate display.

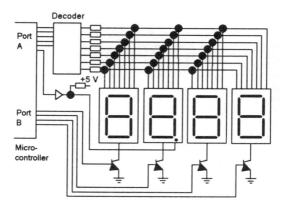

Figure 24.14 Multiplexed display

Further reading: Bolton, W. (2000), *Microprocessor Systems*, Longman, *Design with Microprocessors for Mechanical Engineers*, McGraw-Hill; Vears, R. (1990), *Microprocessor Interfacing*, Butterworth-Heinemann.

Programmable interfaces

With parallel inputs and outputs, microprocessors can have their input/output interfaces provided by discrete components, e.g. a tri-state buffer for an input and a latch for the output. An alternative is to use a *programmable peripheral interface* (PIA) (Figure 24.14). Such a device can be connected directly to the buses of the microprocessor and provide two input/output ports which can be programmed to operate in variety of ways. Port A might, for example, be programmed to be an input and port B an output, or vice versa. Input/output transfers are then controlled by the program.

Further reading: Bolton, W. (2000), *Microprocessor Systems,* Longman; Green, D.C. (1991, 1995) *Data Communication,* Longman; Leventhal, L.A. (1978), *Introduction to Micro-processors,* Prentice Hall; Stiffler, A.K. (1992), *Design with Microprocessors for Mechanical Engineers,* McGraw-Hill; Vears, R. (1990), *Microprocessor Interfacing,* Butterworth-Heinemann.

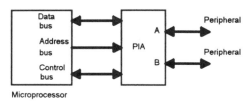

Figure 24.14 PIA

Universal asynchronous receivers/transmitters

The inputs and outputs to microprocessors are parallel transmissions. With parallel transmissions, one line is used to each bit transmitted. With serial transmission, a single line is used to transmit data as sequential bits. The *universal asynchronous receiver/transmitter* (UART) is a programmable interface which can be used with microprocessors to change serial data to parallel data for inputs and parallel data to serial data for outputs (Figure 24.15).

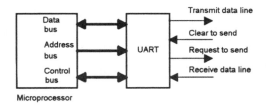

Figure 24.15 UART

Further reading: Bolton, W. (2000), *Microprocessor Systems,* Longman; Green, D.C. (1991, 1995) *Data Communication,*

Longman; Leventhal, L.A. (1978), *Introduction to Micro-processors*, Prentice Hall; Stiffler, A.K. (1992), *Design with Microprocessors for Mechanical Engineers*, McGraw-Hill; Vears, R. (1990), *Microprocessor Interfacing*, Butterworth-Heinemann.

Microcontrollers input/output interfaces

Microcontrollers have parallel input/output interfaces and many also have serial input/output interfaces, i.e. built-in PIAs and UARTs. For example, the Motorola M68HC11 has five ports which can be used as two parallel output ports, one parallel input port and one port which can be programmed as either parallel input or output. By programming it is possible to use some of the pins in these ports as a synchronous series interface and an asynchronous serial interface,

Further reading: Bolton, W. (2000), *Microprocessor Systems*, Longman, Calcut, D. M, Cowan, F. J, Parchizadeh, G. H. (1998), *8051 Microcontrollers*, Arnold; Huang, H-W. (1996), *MC 68HC11 An Introduction*, West Publishing Company; Stewart, J. W, and Mia, K. X. (1999), *The 8051 Microcontroller*, Prentice Hall

Index